John Syer Bristowe, And others

Diseases of the Intestines and Peritoneum

John Syer Bristowe, And others

Diseases of the Intestines and Peritoneum

ISBN/EAN: 9783337035068

Printed in Europe, USA, Canada, Australia, Japan

Cover: Foto ©berggeist007 / pixelio.de

More available books at **www.hansebooks.com**

DISEASES

OF THE

INTESTINES

AND

PERITONEUM.

BY

JOHN SYER BRISTOWE, M.D.

J. R. WARDELL, M.D.,   J. W. BEGBIE, M.D.

S. O. HABERSHON, M.D.

T. B. CURLING, F.R.S., AND W. H. RANSOM, M.D.

―――――――――

NEW YORK

WILLIAM WOOD & COMPANY

1879

CONTENTS.

INTESTINES AND PERITONEUM.

ENTERALGIA.

By John Richard Wardell, M.D., F.R.C.P.

Definition.—Enteralgia is a painful affection of the intestines, of neuralgic character, generally accompanied with constipation and flatus. It may come on gradually in a dull and obtuse manner, but in the great majority of instances its supervention is sudden, and the pain is sharp and violent. It is, correctly speaking, visceral neuralgia, and mostly occurs in neurotic individuals. The common accompaniments of inflammation are absent. The skin is cool, the pulse is not accelerated, and the heart's impulse is rather subdued than augmented. Its attacks are paroxysmal. It shifts its position in the abdomen. It is often a pain reflected by distal disease, but if continuous it may end in inflammation.

Synonyms.—Enteralgia, Tormina, Dolor, Colicus, Colicodynia, Spasmus Intestinorum (*various authors*), Ileus (*Sauvages*), Spasmus Ventriculi (*Wiessner*). Some writers have confounded it with Gastrodynia, or Gastralgia. In the vernacular the affection is identical with Pain of the Intestines, Spasm of the Bowels and Belly-ache, Pain in the Belly, Gripes, and Cholick, or Cholick Colic.

Causes.—The causes of this complaint are to be regarded as those which are *Predisposing* or *Remote*, and those which are *Proximate* or *Exciting*.

Under the head of the first-named may be mentioned the influence of sex, and it is beyond dispute that females are more prone to this affection than males; their greater sensitiveness, and their susceptibility to moral emotions, favor the development of nervous diseases; and the sympathy of the uterus and its appendages, as familiarly known, in marked manner reacts upon the cerebro-spinal and ganglionic systems. The particular temperament of the patient will confer a proneness to, or tend to give an immunity from, this complaint; those who are nervous and melancholic being more liable to it, and those who are leuco-phlegmatic or lymphatic being less susceptible. The condition of asthenia conduces to the production of enteralgia, and a lowered vitalism is often associated with an exaltation of sensibility. The weakness resulting from acute or chronic disease, by depressing the tone of the system in general, and the functional power of the great nervous centres in particular, constitutes a common predisponent, and the morbid action of the nerves proper to some part or parts is not an unusual occurrence. During the convalescence of fever, after visceral inflammation and large losses of blood these attacks are most frequently witnessed. Excessive lactation, by subduing the general strength, often enters as an element into the remote causation; and the same may be said of menorrhagia, the lochia, hemorrhoids, leucorrhœa, and like affections. Long-continued secretions and periodical discharges, by deteriorating and diminishing the vital fluids, are followed by the result in question. Amongst the proximate or exciting causes is to be mentioned the malarial influence, and in tropical countries and aguish districts there is no doubt it often merits the accusation. Atmospheric humidity, low and damp situations, and a naturally cold and wet climate, form endemic conditions which foster the development of neuralgic ailments; and the truth of the converse is unquestionable that in places of greater altitude, and in a purer and drier air they are not so prevalent. When hot and sunny days are followed by frosty nights, the body being suddenly chilled, and thus the blood being determined to the internal organs, these anomalous pains are often produced. Wet clothes and wet feet give rise to the same affection. Mental fatigue, as after long-continued and great intellectual efforts, has by some writers been enumerated. In those persons whose vocations are such as to demand a continued strain of thought, or whose hopes and fears are excited by speculation, as in commercial enterprises, or those whose faculties are stimulated by some career of ambition, in all of whom the nervous functions are brought into great energy of action, these neurotic ailments prevail, sometimes being located in one organ or part, sometimes in another.

There are also proximate causes, which are strictly speaking pathological—which are referable to foregoing and obvious forms of morbid change, especially to those changes which take place in the blood, and which constitute a humoral causation to the nervous phenomena. It has been observed by Simon that central neuralgia arises with the utmost frequency in anæmiated and debilitated persons;[1] and we know how apt it is to follow hemorrhage, and be associated with malnutrition when no primary structural lesion exists. During the latency of the gouty, and in the rheumatic diathesis, when the *materies morbi* of those respective affections has accumulated in the system, before its explosive decomposition has been evinced by local inflammation and excessive secretional evacuation, its presence may be such as to generate that humoral disorder, which first affects the cerebro-spinal and ganglionic centres, and then the nerves proper to visceral organs. In chorea, which is consequent upon some perversion in the development of the blood, caused by the alteration of physical qualities, or the chemical relations of that fluid, or it may be by the absolute generation of some new product, we have ample testimony of the immediate effect produced on the nervous system. And in Bright's disease we are continually presented with examples of the same consequence, caused by the retention of effete and poisonous matters in the circulation. Dr. Todd[2] some time ago pointed out the fact that epilepsy, as associated with this renal affection, is characterized by greater severity in its seizures the longer the interval between the fits, because the irritant

[1] Lectures on General Pathology, Lect X.
[2] Lumleian Lectures *Medical Gazette*, 1849 and 1850.

materials revulsed into the circulation are then in accumulation and act with greater force. The fact that defective blood-development, or its contamination by lesion of the depurative organs, is productive of nervous disorders, is well shown by the administration of suitable remedies. In anæmia and chorea we every day observe the beneficial effects of ferruginous medicines, and see how pains diminish in degree and frequency, and how the disorderly movements of the voluntary muscles become subdued. In hyperæmia, more especially in that form which has been denominated active hyperæmia, pressure upon the nervous filaments gives pain; and although such far more frequently obtains with the solid abdominal organs, yet it doubtless is an element entering into the causation of Enteralgia.

In organic disease of the brain and spinal cord pain is generally reflected to some distant part, and such is the common case in lesion of the last-named organ. In caries of the vertebræ, as I have in repetition observed, the reflected visceral pain has been a constantly recurring sign. Some years ago I saw, at the request of a distinguished provincial surgeon, a lady who for many weeks had been under his care, and whose case he regarded as one of persistent Enteralgia caused by some offending ingesta or some impaction in the bowels. I believed, however, that this pain in the bowels had a more remote origin—that it was spinal. The examination after death revealed vertebral caries and softening of the cord. Sometimes the distal pain can be traced to mechanical injuries of the nerve-centres. We know that in children there is the closest connection between encephalic disease and disorder of the bowels. In primary disease of the solid abdominal viscera, especially in that of the liver and spleen, irritation is not infrequently extended to the intestines; sometimes neuralgic pain of an intermittent or remittent character eventuates; while in active congestion of the liver, or in that sudden distention of the spleen which occurs in periodic fever, intestinal pain is no unusual symptom. The intimate sympathy which subsists between these parts can be well understood when we consider their ganglionic connection.

Amongst the more common causes may be mentioned indigestion and flatulence. When the ingesta have not been properly converted into chyme, but have passed down into the lower bowels only partly disintegrated, they give rise to irregular spasmodic attacks of pain by acting, as it were, like foreign bodies in the canal. In this way shell-fish, dried salt meats, pork, badly cooked food, unripe fruit, crude vegetables, and the like, are followed by the affection. That flatus very often produces Enteralgia is a fact so familiar as scarcely to merit comment; but numbers of the older authors speak of this cause with much emphasis.[1] Wiessner says : " Flatus similiter etiam ventriculum doloribus spasticis afficiunt. Hæc enim toti tractui intestinorum molestissima affectio vel ipsi ventriculo proxime nocet, vel partium distentione stomacho proximarum. Ex hisce imprimis colon transversum, ante inferiorem ventriculi curvaturam extensum, sedem aeri incluso quam maxime incommodam parat."[2] The movement of gases from one part of the intestines to another accounts for the shifting of the pain. Constipation is another and frequent cause of the complaint. Indurated masses of fæces become

[1] Rhodii Obs. Med. cent. iii. Palav. 1657; cent. ii. obs. 70. Lieutand, Hist. Anat. Méd. tome i. p. 7; Paris, 1767. Marchand, Diss. de Cardial. flatul. Argent. 1754. Weikard, Vermischte med. Schriften, Frankf. 1778, b. ii. p. 143.
[2] De Spasmo Ventriculi, p. 13.

impacted in the cæcum, sigmoid flexure, or transverse colon, and attacks of sharp, twisting, rolling pain come on from time to time, and are not permanently relieved until the irritative contents of the gut have been voided. Sometimes a large gall-stone or a concretion is the cause. Morbid secretions, acrid substances, acerb fruits, septic food, such as putrid game and bad cheese, stimulating liquors, and sour drinks are liable to produce Enteralgia. Chemical agents and medicinal compounds are followed by a like result. The sensitive fibriles proper to the lining tunic of the digestive tract, by coming in contact with the fore-named, become irritated, and there may be great pain when the motor nerves are but slightly influenced in their functions. In lead poisoning the intestinal nerves are particularly prone to exaltation of sensibility.

SYMPTOMS.—The mode of accession is generally sudden, the pain being sharp, shooting, or twisting; but in some instances it comes on more gradually, and a rolling or aching of the bowels is described. The affection is in the majority of cases first felt at the umbilicus or in the right iliac fossa. The paroxysms increase in degree and frequency, the intervals from suffering being irregular and of varied duration. The pain, especially in the earlier stage of the attack, alters its position. It is rather relieved than aggravated by pressure. The skin is often cool, the face pale, and the pulse, instead of being accelerated, is rendered slower than natural. In the severer cases the stomach sympathizes, and sickness and vomiting may supervene; and when the malady becomes intensified and the agony excessive, the entire surface is bedewed with a chill, clammy perspiration, the extremities becoming cold and of venous hue, and the general aspect that of collapse. Costiveness is the common accompaniment, and percussion displays an overloaded state of some part of the colon, generally at the cæcum or sigmoid flexure. When flatus is the chief cause, there is intestinal distention, and such notably obtains in the large bowel. On palpation nodulated eminences are felt, which quickly alter in their configuration, and which are caused by the constricted and distended portions of the tube. With the expulsion of the confined gases the patient derives signal and immediate relief, and sometimes the amount evolved is very considerable. The noisy flatulent movements — borborygmi—which are often heard in the canal frequently constitute a marked symptom in hysterical females whose primæ viæ are generally disordered, and whose assimilative functions are imperfectly performed. The attacks of Enteralgia may be intermittent or remittent. Sometimes they terminate with all the rapidity with which they were ushered in. Although, as a rule, Enteralgia is apyrexial in character, yet inflammation sometimes occurs; and then the surface is warmer, the pain more fixed, and the circulation excited. In hysterical women uterine disorder is the usual concomitant, and the enteralgic pain will often be found in association with spinal tenderness. In such instances percussion on the spinal processes should not be omitted, and not infrequently hyperæsthesia of the abdominal surface is a prominent sign. When the subjective symptoms are referable to organic disease, and are evidently reflected, the cerebro-spinal axis and the solid abdominal organs should respectively be examined, and the kind of lesion there existent be as far as possible correctly estimated. This neuralgic pain of the intestines is occasionally seen as a symptom caused by ulceration and congestion of the uterus; and it may come on after the sudden retrocession of cutaneous eruptions and the exanthemata; also, as before remarked, it may follow profuse critical evacuations, the repeated loss of blood by hemorrhoids or other sources of debility.

The symptoms are modified or terminated, or the attacks rendered less recurrent, by the accession of certain morbid conditions taking place in the system. The advent of a powerful diaphoresis, the supervention of diarrhœa, the flow of the catamenia, the lochial discharge, the occurrence of epistaxis, the formation of an abscess, or the return of some long-habituated secretion, are known to exert such influence. A fit of gout, or the development of acute rheumatism, seem on derivative principles to lessen this nerve-pain, and diminish in or remove from the organism those conditions of irritation which particularly affect the cerebro-spinal and ganglionic nerves. It sometimes happens, when the affection comes on in females, that a very large secretion of pale or almost colorless urine is at once succeeded by the mitigation of the attack. In some cases the abdominal and thoracic muscles are spasmodically contracted; there are rigidity of the recti and a loss of motor power in the intercostals, the chest is fixed, and the breathing is oppressed. When the seizures become repeated, they are liable to be characterized by greater severity, the exaltation of the nervous sensibility doubtless becoming augmented by continued irritation. The duration of the symptoms is always most uncertain, as much will depend upon the kind of fundamental cause by which they are produced: it may, however, be regarded as the most usual fact that the more severe the fit the shorter will be its continuance. In children spasmodic pain of the bowels is soon productive of disorder in the digestive functions, and irritation in the alimentary tube is soon followed by the ordinary conditions which characterize infantile convulsions. According to M. Billard, the child cries suddenly and loudly, the face is contracted, the limbs are stiffened, the belly is tender to the touch, there is tympanitic distention, and the attack is often relieved by the expulsion of large quantities of gas *per anum*.[1] The alvine evacuations are generally suspended, and frequently there are vomiting and carpo-pedal contractions; in young infants there is tossing of the arms, the legs are drawn up to the abdomen, and often in the course of time green and offensive stools are voided. Frequently upon investigation it will be found that the mother's milk, or the artificial food which has been given, is the cause.

PATHOLOGY.—In the discussion of this part of the subject, those morbid conditions may first be noticed that consist of impairment of the functions of the bowels, which are characterized by alteration of sensibility, and which are often in association with a lowered state of vitality in the economy; but, as in all functional affections, the real origin of the complaint cannot always be detected, and remains an uncertain inference. It frequently happens that when some irritation of the mucous surface of the bowels is the cause, gases become generated, and painful dilatation of some part or parts of the tube is the consequence. It is probable that a great portion of the gas is secreted from the blood, for the flatus is often produced too quickly and in too great abundance for the presumption that it comes entirely from the decomposition of the ingesta. By this distention of any particular section of the gut there is loss of tone, the contractile power of the muscular coat may be almost or entirely abolished, and the pressure on the sensitive nervous fibriles occasioned by such dilatation will well account for the complaint, because there may be asthenia of this part of the ganglial system in accompaniment with morbid exaltation of sensibility. This irritation of the peripheral nerves, caused by harmful

---

[1] Traité des Maladies des Enfans nouveaux-nés et à la Mamelle. 8vo. Paris, 1828.

ingesta, concretions, vitiated secretions, and the like, affects not only the intestines themselves, but other organs also; and thus it is that the heart and diaphragm are functionally influenced, and hence the depressed circulation and difficult respiratory movement so commonly witnessed in the more violent examples of the ailment. As pain is often to be regarded as the prominent expression of some malady pervading the entire system, and as the functions of any organ may be thus disturbed, it not infrequently occurs that one or other of the viscera is the seat of such disease; and thus it is that in contamination of the blood spasmodic or neuralgic pain of the intestines may result. In saturnine poisoning there is ample illustration of this fact; the poison is transferred into the circulation, the secretions are arrested, and fits of agonizing pain are felt in the bowels: and so it doubtless occurs in those dycrasial affections in which the fluids of the body are degraded by changes more occult, and by the operation of agents less plainly comprehended. In gout and rheumatism, and in Bright's disease, the cerebro-spinal and organic centres are secondarily influenced through debasement of the blood. When gout is retrocedent or rheumatism suppressed, their peccant materials are revulsed upon and irritate some of the internal organs; and in Bright's disease, when the urinary excreta are imperfectly eliminated, the disturbance of the nervous system is exemplified not only in the exaltations of sensibility in the viscera—neuralgia—but in perverted motor function, as evidenced in the reflex action of vomiting and diarrhœa. In simple asthenia, when there is excess of emotional and other motility, as in hysterical females, the kinds of pain in question are readily developed; and if it cannot be said that structural changes do not exist, at least they cannot be indicated.

When organic disease in some cognizable form does constitute the cause of Enteralgia, the examples may be most varied in their locality, degree of objective symptoms, and the kind and amount of their structural alteration. When the primary lesion is in the cerebro-spinal axis, chemical, physical, and mechanical aids to diagnosis are of no avail, and we are compelled to rely upon analogies and subjective representations.[1] One of the most frequent pathologic conditions is that of hyperæmia, which produces irritation and reflex phenomena. According to Brown-Séquard,[2] when the afflux of blood or other morbid change is at the posterior parts of the cerebro-spinal axis, hyperæsthesia is the common result; and it would seem that interruption of continuity of the vasomotor nerves is the fundamental cause of vascular dilatations. In spinal irritation and hysteria reflex visceral pains thus doubtless arise; and it may truly be affirmed that the causes of spasmodic and neuralgic pains are more commonly central than peripheral—in figurative language, they are more frequently referable to the battery than the conducting wires. In positive inflammation of the cerebral and spinal tissues abnormities of function must necessarily arise. The nerves are seldom diseased. Albers and West in only exceptional cases found the vagi morbid on inspections after whooping-cough. Bichat[3] repeatedly examined the nerves in diseases of the viscera without discovering pathologic change. But according to the testimony of various writers, and from my own observations, the nerves are sometimes inflamed, and are subject to other morbid alterations. The neurilemma is the part most prone to inflammation, and

[1] Sieveking, Manual of Patholog. Anat. p. 211.
[2] Lectures on the Central Nervous System, p. 205.
[3] Anatomie générale, i. 225.

Craigie [1] asserts that such condition is a common cause of neuralgic pain. In tetanus and sciatica the entire nerve has been seen red and swollen. The sympathetic ganglia are sometimes diseased; there may be vascularity of the cellular tissue interposed between the elements of the ganglia, and the ganglionic substance has been seen enlarged and indurated. It is also highly presumptive that there may be molecular change in the contents of the ganglionic corpuscles. Neuromatus formations may be the cause of Enteralgia. As I have already remarked, the more common and obvious conditions of visceral disease may produce enteralgic pain, such as thickening of the inner tunics of the bowel, whereby impediment is given to the contents of the tube; an ancient band of lymph giving rise to constriction ; or by some abnormal growth pressing upon the gut. And in the various organic affections to which the solid organs are liable, reflected enteralgic pain is no unusual result. In diseases of the urinary and generative organs of both sexes the kinds of abdominal pain now spoken of not unusually supervene. Ulceration of the uterus and impaction of the ureters sometimes cause Enteralgia.

DIAGNOSIS.—The diagnostic indications of Enteralgia are sudden, darting, plunging, or twisting pains, which come on paroxysmally, the attacks varying in their degree of severity and in their duration. The intervals between the seizures may be almost or altogether free from suffering. The pulse remains unaltered, the surface is cool, and the facial expression is that of pallor and pain. There is moist tongue, no thirst, the bowels are confined, and flatulent distention is the common accompaniment. Pressure on the abdomen relieves rather than augments the pain, and it is not unusual for the patient to press his hands on his belly during the paroxysm as a means of affording relief. The expulsion of gases from the large bowels gives immediate ease; and sometimes the advent of diarrhœa at once cuts short the complaint. In inflammation of the bowels pressure confers pain, the skin is hot and dry, the pulse quick, the face flushed, the secretions and excretions are diminished, the patient cannot turn and twist about in bed as he can in Enteralgia, and the objective symptoms of symptomatic fever are more or less proclaimed. In inflammation the pain is confined to one particular part of the abdomen, and only gradually becomes diffused. In Enteralgia it shifts about with great celerity. In ileus there is vomiting, and at length of fœcal matters; a lump can often be felt, and the suffering, as in inflammation, does not intermit. When this neuralgia of the bowels is from impaction of fœces, palpation and percussion will be our guides; if from concretions or mechanical obstructions, the history of the case and collateral circumstances will conduct to a right decision; and if from irritative secretions, a flux generally supervenes. If reflected by distal disease, as in hepatic, splenic, and renal ailments, those organs should be carefully examined. In neuralgia the pain radiates round to the back, generally at one side. In the passage of renal calculus the pain is in one side; it darts down towards the pubes and thigh, and in the male there is retraction of the testicle. In rheumatism of the abdominal muscles the disease pervades some other part. In hysteria the spine should be examined; and when from this cause, often a copious discharge of colorless urine will give relief. In lead-poisoning there will mostly be dropping of the wrists, and the blue line on the gums.

TREATMENT.—The remedies first indicated are those which are most

---

[1] Pathological Anatomy, 2d edit. p. 380.

likely to abridge and mitigate the sufferings of the paroxysm; and with this view antispasmodics and anodynes may be prescribed, such as opium, chloric æther, henbane, conium, camphor, ammonia, and similar agents. At the same time hot fomentations, sinapisms, terebinthinate epithems, or stimulating and rubefacient liniments, may be employed. The surface should be kept warm and diaphoresis promoted, which can be best accomplished by the patient first putting his feet and legs into hot mustard and water, and then going to bed. The warm bath and sedative enemata are excellent auxiliaries. Sometimes anodyne embrocations, addressed to the spine, do much good. In the more chronic neuralgic affections, I have long been in the habit of prescribing a liniment composed of laudanum, chloroform, the extract of belladonna, and the linimentum camphoræ. The bowels should afterwards be cleared out by mild laxatives, such as castor-oil, the compound rhubarb pill, extract of colocynth in combination with extract of henbane, or the galbanum pill, or the confection of senna. When we believe the fundamental cause to reside in the solid viscera, or in the cerebro-spinal axis or ganglionic centres, our measures should then be addressed to such parts, and our aim be to lessen the general morbid excitability of the nervous system.

ENTERITIS.

By John Syer Bristowe, M.D., F.R.C.P.

The term Enteritis, signifying inflammation of the bowels, is of ancient date, and from the earliest times until now of more or less loose and various application. It has often been applied to a certain group of symptoms irrespective of the conditions under which they may arise, and irrespective even of the presence or absence of actual inflammation, as for instance to strangulated hernia, intestinal stricture, and other forms of obstruction of the bowels; and again the word has often been made to include various specific forms of disease attended with specific intestinal lesions, such for example as enteric fever, tuberculosis, and cancerous infiltration. It is intended in the present article to treat of Enteritis, according to its real meaning, as a simple inflammatory affection; and to eliminate from the subject, as far as possible, all reference to the diseases with which it may be confounded or on which it may supervene.

I. As Affecting the Serous and Muscular Coats.—The intestinal tunics are all of them liable to inflammation either separately or in combination: and the inflammatory process, as it occurs in each, has a tendency to present characteristic peculiarities, and to be associated with special symptoms. Inflammation of the serous coat is of frequent occurrence as a part of general peritonitis, a disease the morbid anatomy and symptoms of which are subsequently described; but, as in the analogous cases afforded by the pleuræ and pericardium, inflammation commencing here spreads rarely, or with difficulty and late, to the subjacent tissues; and hence peritonitis may be considered practically to be as distinct from true inflammation of the bowels, as pleurisy is from pneumonia, or pericarditis from inflammation of the heart. Nevertheless inflammation beginning at the peritoneal surface does occasionally invade the whole thickness of the intestinal walls; and still more frequently, just as pneumonia induces inflammation in the overlying tract of pleura, inflammation of the deeper tissues of the bowel leads to circumscribed inflammation of the investing peritoneum, and to the superaddition of peritonitic symptoms to symptoms previously existing. The structures lying between the serous and mucous tunics, namely, the muscular laminæ with their associated nervous plexuses and connective tissue, are rarely the primary seat of inflammation; occasionally, it is true, in pyæmia and under other exceptional conditions, an abscess forms in them; but they are more frequently involved in the extension of peritoneal inflammation; and still more frequently they become inflamed either by the spread of inflammation from an inflamed mucous membrane, or in consequence of its simultaneous origin in the several intestinal tunics. Inflammation and its results here, in their slighter forms, scarcely reveal themselves to ordi-

nary post-mortem examination, but when more pronounced, are manifested anatomically by congestion and effusion of serum, lymph, pus, or
blood. The symptoms which they induce are in the first instance probably spasmodic contraction of the muscular fibres, subsequently loss of
power or complete paralysis.

II. As AFFECTING THE MUCOUS MEMBRANE.—Inflammation as a primary and characteristic affection occurs far more frequently in the mucous
membrane than in the coats external to it; and it occurs here in forms
which vary considerably according to its cause, the constitutional conditions under which it arises or with which it is associated, and its degree
of intensity.

(a) *Catarrhal Inflammation.*—The slightest and simplest form of
inflammation is usually termed *catarrhal.* This may be produced by the
local action of irritating ingesta, or by the influence of those external
conditions which are known to be the agents in setting up the same kind
of inflammation in other parts; and it is believed by some to attend generally scarlatina and other specific fevers.[1] Young children, particularly
during the period of teething, seem specially liable to it. It is characterized by congestion, tumefaction, softening and dryness of the mucous
membrane, followed speedily by the secretion, often in considerable abundance, of mucus, which is ropy or watery, irritating, and sometimes mixed
with blood. It sometimes affects the lower bowel only, producing mild
dysenteric symptoms; but frequently it commences in the upper bowel,
or in the stomach, and spreading thence downwards gradually traverses
the whole of the intestinal canal, causing in its progress more or less uneasiness, aching and griping, attended frequently with nausea and sickness while it is still high up, with diarrhœa and expulsive pains and
efforts when it reaches the large intestine. The tongue is generally more
or less furred and dry, the breath offensive, and the appetite impaired;
but these symptoms vary, and are often absent, especially when the large
intestine alone is affected. Some degree of general febrile disturbance,
indicated by heat and dryness of skin with sense of chilliness, increased
frequency of pulse, lassitude and headache, is usually attendant on the
local disorder. In children, in whom inflammatory affection of the gastro-
intestinal mucous membrane is sometimes associated with aphtha, the
disease not infrequently produces serious results and death, either from
the debility which follows persistent diarrhœa and vomiting, or from the
supervention of cerebral complications, such as convulsions or coma.
There can be no doubt that a large number of cases of gastro-intestinal
disturbance and of diarrhœa are due to catarrhal inflammation; yet the
existence of such inflammation is more a matter of inference from symptoms than of direct observation upon the condition of the mucous membrane. For the latter can only be examined after death, at which time
congestion and other indications of superficial and slight inflammation
have for the most part disappeared, or are lost in post-mortem changes.

(b) *Croupous Inflammation.*—The designation "croupous" (diphtheritic or membranous) inflammation is given to those cases in which
the mucous surface becomes covered to a greater or less extent with a
more or less adherent membranous film consisting of corpuscular elements
cemented together by a coagulable exudation, and prolonged for the
most part by rootlets from its under-surface into the Lieberkuhnian folli-

---

[1] See Dr. Fenwick on "The Condition of Stomach and Intestines in Scarlet
Fever," Med. Chir. Trans. vol. xlvii,

cles. This affection, which is far from uncommon, may sometimes doubtless be regarded as the expression of some specific form of inflammation;
certainly many believe (and I am one of them) that it is a common feature in the early stage of dysentery; at the same time it frequently occurs
quite independently of all infectious or malarious influence. It undoubtedly indicates greater intensity of inflammation than mere catarrhal inflammation; there is generally much greater congestion and thickening
of mucous membrane, and not unfrequently hemorrhage, suppuration, or
gangrene. Croupous inflammation is often met with in the large intestine in scattered patches, which are sometimes linear, sometimes irregularly polygonal or stellate, and occupy for the most part the prominent
ridges of the mucous membrane, more especially the edges of the inter-
saccular constrictions. In some cases, still chiefly occupying the more
prominent parts, it forms a coarse, irregular network extending over large
tracts of surface; in other cases it forms uniform patches of considerable
extent. It is less common in the small intestines, but may be found in
them affecting the free edges of the valvulæ conniventes, or spread over
a large area. It is sometimes met with on the surface of tracts of cancerous infiltration which are on the eve of ulcerating. It may be added
here that cases sometimes come under observation in which patients pass
per anum shreds of false membrane, or even membranous casts of the
bowel, of soft texture, various thickness, and of a dirty greenish or
brownish hue. This discharge is generally, if not always, a consequence
of dysenteric ulceration. The symptoms which attend croupous inflammation are not special; they vary, according to circumstances, on the one
hand between those of diarrhœa and dysentery, and on the other hand
between those of mere colic and of typical enteritis. The patchy form,
indeed, so common in the large intestine, is often overlooked during life,
from the fact that it occurs as a complication in the later stages of many
grave disorders, as, for example, acute pneumonia, Bright's disease, cirrhosis of the liver, and cerebral affections.

(c) *Chronic Inflammation and Degeneration.*—Both catarrhal and
croupous inflammations, in their slighter degrees, generally, and for the
most part speedily, undergo resolution. Sometimes, however, they end
in ulceration; an event which, with its consequences, is fully considered
further on. And sometimes they lead to persistent modifications of the
mucous membrane which are often included in the term "chronic inflammation." These consist generally in slight condensation and hardening
of the mucous tissue, more or less distinct congestion, or black pigmentary
deposit in the villi and interfollicular spaces, some degree of atrophy of
the Lieberkuhnian follicles, and granular or fatty degeneration of their
epithelial contents, together with an analogous condition, more or less
pronounced, of the epithelium of the mucous surface generally. The solitary and agminated glands are sometimes atrophied, sometimes larger and
more obvious than natural. The changes indeed are chiefly changes of
degeneration; and in that sense, as probably also clinically, are related to
the lardaceous degeneration which occasionally happens in persons laboring under chronic tuberculosis, bone disease attended with suppuration,
and secondary syphilis.[1] Lardaceous degeneration occurs later in the
bowel than in the liver, spleen, and kidneys; it is found chiefly in the
lower part of the ileum and in the large intestine; it affects in the first

[1] See a good account of lardaceous degeneration, M. Hayem, quoted in New Sydenham Society's Biennial Retrospect of Medicine and Surgery, for 1865-6, p. 176.

instance the small arteries and capillaries around, and in the solitary and agminated glands, which bodies become swollen; and then gradually tends to involve the whole thickness of the intestinal wall, the muscular fibres and other tissues becoming finally infiltrated. The bowel thus becomes thickened, and at the same time harder than natural; and often in the later stages erosion of the affected glands occurs, leading in Peyer's patches to a reticulated condition of surface. The above chronic affections of the mucous membrane are generally associated with diseased conditions of other organs, to which indeed they are secondary; and not infrequently the stomach is at the same time the seat of some chronic morbid process. The presence of these complications, and the fact that clinically ulceration of the bowels, together with tubercular and other morbid processes, passes in a large number of cases for chronic inflammation, render it difficult to isolate the clinical phenomena due specially to the bowel affections now under consideration. They doubtless vary greatly; but may be briefly summarized as combining in various proportions, both relatively and posi- tively, imperfect digestion of the alimentary matters received into the in- testine, excessive secretion of more or less watery mucus, increased peri- staltic movements with griping pains, looseness of bowels with discharge of watery, or yeasty, or otherwise unhealthy and offensive evacuations, and innutrition from the imperfect absorption of food.

III. As AFFECTING THE WHOLE THICKNESS OF THE BOWEL.—By the older writers generally, and for the most part also by those of more recent times, the simple unqualified name "Enteritis" has been used to signify a special group of symptoms associated with the presence of a more or less extensive tract of intensely inflamed bowel. The affection here referred to is termed by Cullen phlegmonous enteritis, in contradistinction to the milder varieties of inflammation, affecting the mucous membrane only, which he included under the name of erythematous enteritis.

The symptoms which are supposed to characterize this form of enteritis may creep on insidiously or show themselves in sudden intensity, and con- sist mainly, in the earlier stages, in more or less severe abdominal pain (resembling in its character and in its increase by pressure and by move- ment the pain of peritonitis, but differing from it in being associated with colic), obstinate constipation, nausea and vomiting (occurring both after and independently of the ingestion of food), and marked febrile disturb- ance; and subsequently (supposing the case to be going on unfavorably) in the gradual supervention of tympanitis, attended, for the most part, with diminution or even total cessation of abdominal pain and tenderness, with still persistent constipation and vomiting (the vomited matters becom- ing opaque, brown, and fœtid, if not actually fæcal), with hiccough fre- quently, and with collapse (indicated by extreme feebleness of pulse, cold- ness and dampness of the surface, especially in the extremities), and finally death from asthenia. The morbid changes which may be looked for after death are such as are produced by intense inflammation of a limited tract of intestine. The affected part, which is mostly in the small intestine, and which may vary in length from an inch or two to one or two feet or more, is as a rule much dilated; its serous surface presents a general dusky red, or slate, or purplish-black color, due to the condition of the parts internal to it; it is marked, too, by lines or patches of more or less intense superficial congestion, may present blotches of sub-serous extravasation, and is often covered more or less with adherent lymph; its mucous and sub-mucous tissues are mostly somewhat thickened and softened, some- times only moderately congested, but presenting spots and streaks of extravasation, sometimes black from combined congestion and extravasation, sometimes pale and infiltrated with lymph or pus, sometimes distinctly gangrenous; and its middle coat, sharing in these changes, is also more or less swollen and soft, and congested or œdematous, or the seat of some form of inflammatory exudation. The inflamed tract usually presents fairly well-defined limits, terminating abruptly below in pale and healthy but contracted and nearly empty bowel, above in bowel which may also be healthy, but is dilated like the diseased portion and filled like it with fæcal contents. The diseased intestine contains frequently in addition to simply fæcal matters more or less sanguineous exudation; and traces of the same exudation may often be discovered in the contracted bowel below.

Now, the above phenomena are by no means infrequently met with; they are the common accompaniments of strangulated hernia and of intussusception; they are present in those cases in which, as is supposed, the sigmoid flexure or some other loop of bowel becomes twisted on itself and thus strangulated; they supervene whenever a gall-stone or other foreign body of sufficient size becomes fixed in its passage along the intestine; they occur sometimes also as a late event in stricture, or in those cases in which the bowel becomes constricted by bands of lymph; they are sometimes developed as a result of the extension of inflammation, either from peritoneum or from an intestinal ulcer; and very rarely indeed they originate idiopathically, that is to say from such general causes as produce idiopathic peritonitis, idiopathic pneumonia, and the like. Enteritis, therefore, is a disease which is almost always complicated with some other grave lesion, on which indeed it depends, and which modifies alike its symptoms and its progress.

But even in the uncomplicated form of the disease, which is alone now under consideration, the symptoms are liable to considerable variety; the variations depending mainly on the degree of intensity of the inflammation and its extent, and on the situation of the affected portion of bowel. Indeed, the two principal factors in producing the characteristic symptoms of enteritis are inflammation, on which depend the various febrile phenomena, and paralysis of the inflamed portion of bowel, which permits of its passive dilatation by the accumulation of contents, opposes a more or less complete bar to their transit, and thus induces on the one hand constipation, on the other vomiting.

The most important practical distinction between colic and enteritis is, according to most authors, the absence of febrile symptoms in the former disease, their presence in the latter. And no doubt in most cases of enteritis febrile symptoms manifest themselves in a marked degree, at least in the earlier stages of the malady. Heat of skin, rigors, quickness and hardness of pulse, not infrequently mark the onset of the attack; but it is a mistake to suppose they are always present, or at all events readily perceptible, for in many cases no rigors are experienced, and in some there is little or no acceleration of pulse until towards the close of life, and no more heat of surface than may attend, and often does attend, the gripings of ordinary colic. There is mostly some dryness and clamminess of mouth, if not absolute thirst; and the tongue, which is occasionally pretty clean at the beginning, becomes generally soon thickly coated and ultimately dry. Another feature of enteritis upon which much reliance is placed is the association of the abdominal pain and tenderness of peritonitis with the tormina of colic. Pain and tenderness are certainly present in most cases, at least in the beginning, and in dependence upon them the dorsal decubitus, so characteristic of peritoneal inflammation. But these symptoms vary greatly; sometimes they are intensely severe, sometimes they are from first to last scarcely appreciable, and generally they subside in the progress of the case. It can readily be understood that when the peritoneal surface is largely involved, the pain and tenderness will generally be proportionably severe; that when an extensive length of bowel is affected, there will be correspondingly extensive uneasiness and tenderness; and that when, as sometimes happens, the serous surface is not inflamed, or when the affected portion of bowel is small, the pain and tenderness may be not only limited in extent, but no greater than one finds them in colic or in simple ulceration of the mucous membrane. It is worth while to remark, that limited pain and tenderness are very commonly referred to the region of the umbilicus. Tormina are often at the onset very agonizing, being then probably due in some measure to the spasmodic movements of the inflamed bowel; but they continue even after paralysis has become established, in consequence of the violent but ineffective efforts of the bowel above the seat of disease to overcome the impediment which the disease produces. But tormina are sometimes scarcely recognizable, and frequently, like pain, cease comparatively early. Constipation and vomiting are among the most essential symptoms of enteritis. In the uncomplicated affection the impediment to the action of the bowel is due simply to the presence of a paralyzed and inactive zone of greater or less breadth between an upper and a lower length of healthy bowel; it is no necessary part of the disease, therefore, that the outbreak of acute symptoms shall have been preceded by constipation, or even that after the disease has become established the portion of the bowel below the inflamed part shall not empty itself; and, it may be added, that in a variable degree the contents even of the inflamed gut may slip or be squeezed onwards into the healthy tube beyond, and that even calomel, and such other purgatives as act rather through the system than directly, may produce to some extent their characteristic effects. But it is nevertheless a fact that the inflamed bowel is really a substantial impediment, that there is therefore during the progress of the disease marked constipation, and that purgatives as a rule produce no purgative effect. Vomiting may occur in colic, in diarrhœa, in simple peritonitis, and in many other conditions as a mere sympathetic affection; and sympathy has probably some share in its production even in enteritis, at least at the commencement. But ultimately the vomiting here is due directly, like the constipation, to intestinal obstruction. In the first instance, no matter where the obstruction or what the immediate cause of vomiting, the vomited matters are merely the secretions of the stomach mixed with alimentary substances; but soon bile becomes mixed with these; and before long glairy mucus and bile alone are discharged. Then the eructations become fœtid; and soon the fluid brought up gets turbid and brownish, and by degrees comes to resemble the contents of the lower part of the small intestine, but it becomes fœtid also, and sometimes much more fœtid than the contents of a healthy bowel ever are, the fœtor being caused partly by decomposition of the fæcal matters, partly, as in dysentery, by the discharges taking place from a gangrenous or otherwise diseased mucous surface. This vomiting of the contents of the intestines is, as Dr. Brinton has well explained, not due to inversion of peristaltic action; but is the result of the gradual accumulation of matters in the bowel above the seat of disease, of their mixture gradually effected by the normally-directed peristaltic movements of the bowel, and of their escape into the stomach partly by simple overflow, induced sometimes by mere change of posture, partly by the pressure exerted on the distended bowel by the surrounding viscera, and by the muscular walls of the abdomen. The fœtid matters which thus reach the stomach often, towards the close of life particularly, escape from it into the mouth by mere regurgitation. Tympanitis is probably in no case wholly wanting; in an early stage it may be, and perhaps usually is, absent or but little marked; ere long, however, the abdomen begins to enlarge, and generally as the case progresses becomes greatly distended, tense, and drum-like. This condition is of course mainly due to the distention by fæcal contents and flatus of the portion of intestinal tube which is inflamed and of that which is above it, but now and then it is connected with rupture of the distended intestine and escape of gas into the peritoneal cavity—an accident, it need scarcely be said, of fatal augury. The pulse at the beginning is, as has been already remarked, often accelerated and hard, but it varies greatly in different cases, both in frequency, volume, and strength, and sometimes retains pretty nearly its ordinary healthy character throughout at least the earlier stages of the disease. As the fatal issue, however, approaches, it becomes more and more feeble, and sometimes at length wholly imperceptible at the wrist; it generally becomes then also quicker, sometimes slower, and not infrequently irregular. The temperature of the skin is usually in the first instance more or less elevated, and its surface dry; but even then perspirations are apt to break out, especially during the paroxysms of colicky pain: subsequently, however, the temperature falls, the extremities and face become cold and pale, or livid, with sometimes a faint tinge of jaundice, and all parts of the surface bathed in profuse cold perspiration. The expression of the patient is generally indicative of anxiety and distress, and it has often been noted that, towards the close of life, the face becomes pinched and shrivelled, and assumes an unnatural aspect of old age. He generally retains his senses throughout his illness, and even up to the moment of death: but this event is often preceded by a period of quiescence or lethargy, and occasionally by slight rambling and almost complete unconsciousness. It may be added here, that there is generally in enteritis more or less complete suppression of urine, a phenomenon which has been variously interpreted, but which is probably due, as Mr. Sedgwick[1] argues, to the influence of the abdominal sympathetic system.

Enteritis, in that intense form of it which has been now described, is undoubtedly a very fatal, and indeed very rapidly fatal, malady. It is so difficult, however, practically to isolate the comparatively few cases in which it forms the primary and sole disease from the many in which it supervenes as a complication of some pre-existing graver lesion, that the former scarcely admit of statistical examination. As respects the duration, however, of fatal cases, it may be asserted that it rarely exceeds a week, and that it may be as short as twenty-four or thirty-six hours.

IV. TREATMENT.—It seems scarcely necessary to discuss here the treatment of simple catarrhal and croupous and chronic inflammation of the bowels; these inflammations, indeed, are so intimately connected, on the one hand with inflammatory conditions of the stomach, on the other with dysentery and diarrhœa, which have all been elsewhere described at length, that the reader may be safely referred to the articles relating to those diseases for the principles and details of treatment applicable to the inflammations now in question.

---

[1] Med.-Chir. Trans. vol. li.

In reference to the treatment of the more severe forms of enteritis, two main principles seem now to be fairly well-established : they are, first, to relieve pain, and prevent, so far as may be, all movements of the bowels, by means of opium ; secondly, to avoid every attempt (at least until all grave symptoms have ceased) to force the bowels by the administration of purgatives. It has been shown quite conclusively, principally by the experience derived from the after-treatment of strangulated hernia, that it is always dangerous to endeavor to propel fæcal matters through an enteritic length of bowel, that in most cases the effort is useless so far as their effectual propulsion is concerned, while, by the augmented muscular and excretory action which is thus produced in the bowel above, the diseased tract below becomes more and more distended, almost certainly more and more softened, congested, and inflamed, not infrequently becomes ruptured, and at the very least has its progress towards recovery delayed. Besides which, purgatives tend greatly to increase pain, and vomiting, and general distress. And, indeed, when one considers the great length of time during which constipation may continue with little or no influence on the general health, how long patients with impassable stricture of the bowel manage often to survive, it must be obvious that the constipation of a disease of so short duration as enteritis is not of itself a grave source of danger. Clearly, if the patient is to get well, his recovery must in the first instance be dependent on the recovery by the diseased bowel of its healthy tone, and capability of peristaltic action : and to this end our efforts must be directed. But experience shows us that we have little or no power to arrest internal inflammation, unless it be indirectly by promoting the quiescence of parts, and by relieving pain and irritation; and, for these purposes, opium, in large and frequent doses, is generally our most valuable agent. No absolute rule can be laid down with regard to the quantity of opium which should be given for a dose, or to the frequency with which the dose should be repeated; the patient should, however, be got well under the influence of the drug, and should be kept under its influence. But the constant vomiting and the distention of the bowels above the seat of disease, form a serious, if not fatal impediment to the absorption of opium received into the stomach; what is swallowed may be wholly vomited, or, if retained, very partially or not at all received into the system. If therefore it be thought right to administer opium by the mouth, it should be given in the form least liable to provoke, or to be rejected by, vomiting; but it is certainly best to administer it in the form of suppository or enema, or to inject it subcutaneously. But, no doubt, it is generally desirable, and even necessary, to associate with the use of opium other details of treatment. The question of the abstraction of blood, formerly so largely employed in the treatment of internal inflammations, is not unlikely to arise; and it must be acknowledged that there are cases in the early stage of which removal of blood may be advantageous. When, at the commencement of enteritis, the symptoms of peritoneal inflammation are strongly pronounced, there is no doubt that the application of twenty, thirty, or more leeches to the surface of the abdomen is generally followed by great and immediate relief, if not by actual benefit. Doubtless, the removal of blood from the arm would be at least equally beneficial; and in cases in which, at the same stage, peritonitic symptoms are less distinct, but in which there is high fever, I should not hesitate to have phlebotomy performed. Warm but light applications to the surface of the belly generally soothe, even if they produce no further beneficial effect; and sometimes mustard-plasters, and similar mild counter-irritants, give relief. In the same way, enemata of warm water or of warm gruel are at times useful. There are few symptoms more distressing to the patient than the persistent nausea and vomiting from which he suffers, and few therefore which we feel more anxious to relieve; but there are none which, at all events at certain stages of the disease, are less under the influence of direct treatment. At any early period, when these symptoms are merely sympathetic, ice, hydrocyanic acid, alkalies, lime-water, bismuth, carminatives, and other remedial agents, may no doubt restrain them to some extent; and again, when the disease has begun to take a favorable course, they subside naturally, without any special treatment; but when the vomiting is simply the result of over-distention of the stomach and bowels, to which over-distention there is no other channel of relief, medicine ceases to have any power over it. The extreme prostration which so early manifests itself, is a strong indication of the need of food and stimulants; but how can they be administered with even a chance of benefit? Their exhibition by the mouth tends to promote sickness, tends also to add to the distention of the already too much distended stomach and bowels, while probably, from various causes, little or nothing of them becomes absorbed. It is obvious, indeed, as is insisted on by Dr. Brinton,[1] that alimentary matters, if given by the mouth, should only be given in very small quantities, and in a form suitable for their ready appropriation by the system. They may, however, be given in much larger quantities, and with none of the above ill effects, and also with a much greater chance of benefit, in the form of enemata. It is not intended by the above remarks to discourage all attempts to restrain sickness, or to supply stimulants or food; for there are cases which seem hopeless, in which, nevertheless, the bowel is recovering, and in which the alternative of life or death depends upon the judicious use of remedies and of regimen; but only to discourage persistence in lines of treatment when their effect on the patient, and the progress of the case, prove their inutility or harmfulness.

---

[1] Intestinal Obstruction. 1867.

2

OBSTRUCTION OF THE BOWELS.

By John Syer Bristowe, M.D., F.R.C.P.

The affections which are here to be treated of present many features in common with enteritis, and their description is not infrequently included in the description of that disease. Actual enteritis does indeed occur at some period or another in the course of most of them; but their special claim to form a group by themselves consists in the fact of the existence in all of them of some mechanical impediment to the transmission of the contents of the bowels, in connection with which enteritis is apt to, but does not in all cases necessarily, supervene. They are: 1st, constipation; 2d, stricture; 3d, compression and traction of the bowel; 4th, internal strangulation; 5th, impaction of foreign bodies; and 6th, intussusception.

I. Constipation.—(a) *Pathology and Symptoms.*—Constipation not only forms a more or less essential element in the history of all the affections just enumerated, but of itself induces occasionally insuperable obstruction; and on both of these grounds demands some brief consideration here. Prolonged retention of fæces is within certain limits, of such common occurrence, and is attended with so little inconvenience, that it scarcely deserves in a large number of cases to be regarded as an abnormal condition. It may doubtless be accepted as a general rule, that persons enjoying robust health, and unimpeded in the regular performance of their various functions, have an alvine evacuation at least once daily. Yet many who are apparently equally healthy have their bowels relieved habitually every two or three days only, or even but once in a week or fortnight. Cases indeed are not altogether rare in which some degree of good health has been maintained for many years although fæcal evacuations have during that time occurred only at intervals of six weeks or two months. In the case of a lady recorded by Dr. Robert Williams,[1] in whom habitual constipation appears to have been augmented by the constant use of large quantities of opium, the bowels were frequently confined for six weeks together, and during one year of her life there were only four evacuations at intervals of three months. It must not be forgotten, however, that that degree of constipation which is habitual with one man, and in him compatible with perfect health, may be and often is, a source of discomfort, if not of positive illness, to another man in whom its occurrence is exceptional. Thus, to most persons whose daily habits in this respect are regular, the retention of fæces for two or three days is apt to produce not only local uneasiness, such as fulness, heat, tendency to piles and flatulence, but also some degree of general constitutional disturbance indicated

---

[1] Dr. Burne on Habitual Constipation, quoted by Mr. Pollock in Holmes's "System of Surgery," vol. iv.

by headache, foul breath, loss of appetite, and dyspeptic symptoms, and not unfrequently terminates with more or less tenesmus, or even slight dysenteric diarrhœa. But even in cases in which, from long habit, constipation has come to be regarded as the normal condition of things, some of the above specified discomforts do actually for the most part coexist in some degree with it, but having become, like the constipation, habitual, cease to be observed, or at all events become tolerable. It is easy indeed to see that constipation must tend to produce various inconvenient results: the retention of a mass from which gaseous matters are being constantly evolved, is necessarily productive of colicky pains and imperative desire to discharge flatus; the constant pressure of a hard mass immediately above the anal outlet causes not only congestion of the mucous membrane of the part, but retardation of blood in the hæmorrhoidal veins, and ultimately piles,—it produces also not infrequently some degree of uneasiness in connection with the genito-urinary organs; lastly, when defæcation occurs, the expulsion of the fæces is apt in consequence of their bulk and hardness and dryness, not only to be attended with very considerable pain, and perhaps some loss of blood, but to be followed by prolonged burning or aching, and (as has been already pointed out) by more or less dysenteric inflammation.

But much-prolonged constipation leads sometimes to other and far more serious results, namely to dilatation and hypertrophy of the intestine, ulceration of its mucous surface, and perforation of its walls with extravasation of fæcal matters into the peritoneal cavity. The dilatation is sometimes so great, that the colon measures from nine to ten or even twelve inches in circumference. It begins at a distance of one or two inches from the anus (which seems spasmodically contracted) and occupies more or less of the remainder, sometimes the whole length of the large intestine; in which latter case the chief distention is observed in the rectum, sigmoid flexure, and cæcum. Hypertrophy of the muscular coat, which always accompanies dilatation, is general, but most marked in the sigmoid flexure and upper part of the rectum, where the thickness may be ¼ inch or more. When ulceration takes place, it is perhaps partly due to yielding of the mucous membrane from over-distention, partly to the constant irritation kept up by the fæcal mass within. Perforation may ensue, either while the constipation remains unrelieved, and then either through the progress of ulceration or by laceration; or after the bowel has been emptied, in consequence of the continuance of ulceration. Enormous quantities of fæcal matter are sometimes removed from patients suffering from aggravated constipation; in Dr. Williams' case above referred to, numerous round lumps, each the size of a large fœtal head, were passed at a time, and often in sufficient numbers to fill a common-sized pail.

I recollect two fatal cases which strikingly illustrate some of the observations which have just been made. The first was that of a little girl, eight years old, whom I saw casually only during life, and of whose history I obtained after her death some not very perfect details. She had long suffered from tendency to constipation; and it was stated that she had occasionally gone as long as three weeks without passing an evacuation. At the time of her admission into the hospital there had been no relief to the bowels for seven weeks. She was then pale and thin, had a large tense belly, without pain or tenderness, a clean tongue, and a poor appetite. She had a " strumous " look, and was supposed, I believe, to be suffering from abdominal tubercle. She became gradually more and more emaciated and anxious-looking, while the belly grew larger and more tense. She never had any distinct abdominal tenderness, but suffered at times from colicky pains, and often (especially towards the close of life) complained that she was so full that she felt as if she should burst. During the last week or two the tongue became somewhat foul, and she had frequent vomiting, but never of stercoraceous matter. She passed but little urine, and that was high-colored. She sank gradually from exhaustion, and died exactly three weeks after admission. Amongst other kinds of treatment adopted was the use of purgative medicines and of purgative injections ; and the medical man in attendance on her was led to believe that they had acted. There is no doubt, however, from subsequent inquiries, as well as from what was observed after death, that he was deceived. At the post-mortem examination, the form of the distended intestines was distinctly impressed on the tense and thin abdominal walls, and on opening the abdomen the enormously enlarged colon was at first alone visible. The distention began at the cæcum and extended to within two inches of the anus, where it ceased abruptly. In the greater part of its extent, the bowel measured from nine to ten and a half inches in circumference, the greatest amount of distention being manifested in the sigmoid flexure. The muscular walls were hypertrophied from the ascending colon to the lower end of the sigmoid flexure; and in the latter situation (where the hypertrophy was greatest) they measured $\frac{1}{8}$ inch in thickness. The mucous membrane seemed healthy in the greater part of its extent, but it presented some congestion here and there, and at distant intervals large patches in which there were groups of small circular shallow ulcers. The bowel contained no flatus, but was completely full of thick, semi-solid, olive-green-colored fæces. These were more solid in the rectum than elsewhere, and immediately above the anus formed an indurated conical lump. The small intestines were also considerably distended, though much less so than the larger bowel, and were filled throughout with semi-fluid olive-green-colored contents. The stomach was small and healthy and empty. There was no other disease. There can be no doubt that the death of the child was due to the neglect of simple constipation, that the indurated fæcal lump above the anal orifice had formed a plug which the bowel had been unable to expel, and which the accumulation of more and more fæces above and around it had served only to fix more securely. That the bowel had striven to expel its contents was shown by the hypertrophied condition of its muscular coat. A very similar case is recorded by Mr. Gay ; [1] but there the nature of the case was recognized, the rectum was relieved by mechanical means, and the child was saved. The second case referred to above was that of a young man, aged 24, who also had been the subject of habitual constipation ; and who on one occasion, after the persistence of constipation for an unusually long period, was attacked with diarrhœa, which lasted about six weeks, and was then followed by sudden peritonitis, of which he died. There was found after death inflammation of the peritoneum, due to a perforation in the transverse colon, great dilatation and thickening, yet almost complete emptiness of the whole length of the large intestine, and just the same kind of ulceration of the mucous membrane in patches as that described above. It was in one of these patches that perforation had taken place. Here, as in the former case, it is obvious that long continued constipation had caused permanent thickening and dilatation of the large intestine, and ulceration of its mucous surface ; but here, additionally, after the relief of the constipation, the ulceration had provoked and maintained a condition of diarrhœa, and ultimately caused perforation.

Constipation, in the sense in which the word is here employed, is probably always due to retention of fæces in the lower part of the large intestine, either from failure to respond to the desire for defæcation when the desire presents itself, or from sluggish action on the part of the lower bowel. It is very rare indeed, if there be no actual obstruction, that the contents of the alimentary canal do not pass along the whole length of the small intestine, and even along the colon, at a tolerably uniform rate; at all events, any actual arrest of their transmission, unless it be owing ,to the presence amongst them of some massive foreign body, is probably never met with except occasionally in the cæcum and sigmoid flexure.

Constipation is due to a variety of causes, and occurs under numerous different conditions, which it is scarcely necessary to enumerate here, far less to consider in detail. It is frequently caused temporarily by change of diet, scene, or habits, among which latter may be included anything which interferes with the regular performance of defæcation ; it happens commonly in various kinds of disease, and it occurs in a chronic form in chlorotic or dyspeptic girls and young women, and also in men and women (especially the latter) of sedentary habits or of sluggish constitution. It occurs too, often perhaps as the result of habit, in persons, young and old, in whom no special cause for it can be recognized ; and indeed, in many of the more remarkable cases that come under observation, it is quite impossible to assign a definite cause for it. Among local conditions which may be supposed to operate in a greater or less degree in the above cases, are : first, modifications in the character of the fæces such as we see in diabetes, where, owing to the rapid escape of fluid by the kidneys, they become preternaturally dry, and proportionately diminished in bulk ; second, sluggishness on the part of the rectum ; and third, debility of the same part which may be primary, and due in the first instance to simple thinning and weakening of the muscular fibres, and which probably occurs virtually in all cases of long-continued constipation, when the bowel has become dilated, and on that account (even if the muscular coat be hypertrophied) less competent to contract efficiently on its contents.

(b) The Treatment of constipation must be made to depend more or less upon its cause, on its antecedents, and on its effects. Where it is a mere temporary matter, depending on accidental circumstances, or arising in the course of acute diseases, its treatment is simple enough, and needs no description here. When it has become a chronic affection, its causes should be investigated, and as far as possible obviated; and it may be necessary to employ habitually mild aloetic or other purgatives, or enemata. Sometimes the application of galvanism to the surface of the abdomen, or to the abdomen and anus, is efficacious. But iron and other tonics also are frequently of advantage; and strychnia is by many believed to be of great value. In cases in which the rectum becomes filled with a hard immovable mass, and the bowel above distended in consequence with accumulated contents, the evacuation of the rectum by mechanical means becomes essential. This may be effected sometimes by the use of the finger or of a spoon, or some such instrument; sometimes by the employment of copious enemata administered in the ordinary way; or, better still (as in Mr. Gay's case), by directing a forcible stream of warm water, conducted from a height by means of a tube, into the rectum, allowing it to play upon the fæcal mass for half an hour or so at a time, and thus to cause its disintegration, and either effect or facilitate its removal.

II. STRICTURE.—By this term is meant a circumscribed diminution in the calibre of the bowel, due either to contraction of the mucous and submucous tissues (the consequence usually of ulceration), or to some deposit or growth involving the general thickness of the walls and encroaching on the canal, or to some spasmodic action of the circular muscular fibres. It is occasionally the result of malformation.

(a) *Pathology.*—Congenital stricture, though in some of its forms by no means rare, is an affection the treatment of which belongs almost exclusively to the surgeon, and one, therefore, that needs little more than incidental mention here. It is limited, indeed, with few exceptions, to the lower extremity of the bowel—the rectum and the anus—one or both of which parts may be found at birth to be impervious or absent, or reduced to a mere fistulous canal or orifice, while, in addition, the lower end of the fully-dilated bowel above occasionally communicates with the vagina in the female, or with the bladder or urethra in the male. Very much more rarely, congenital stricture is met with in the duodenum, at or above the point at which the common bile duct discharges itself. Two cases of this kind are recorded in the twelfth volume of the "Pathological Society's Transactions," one by Dr. Wilks, the other by Dr. G. Buchanan. In both, a kind of membranous septum existed at the point referred to, and the portion of the duodenum above was thickened and dilated, forming a mere prolongation of the pyloric end of the stomach. In Dr. Wilks's case the bile duct opened immediately below the septum, which was impervious; and the child died at the end of thirty-eight hours, its death being preceded by vomiting and convulsions. In Dr. Buchanan's case, the duct opened on the under-surface of the septum, the septum presented a minute central orifice, and the child, a girl, lived eighteen months. According to the history, she was apparently quite well up to within a month of her death, probably because (as is supposed) she had hitherto been fed only from the breast and with milk. She appears during the last month of life to have suffered from constant vomiting, great restlessness and uneasiness or pain, together with (during the earlier part of that time) frequent convulsions. It may be added, that in this case, where the parts were examined with much minuteness, the septum was ascertained to consist of a duplicature of mucous membrane, not unlike an enlarged valvula connivens, inclosing a few scattered muscular fibres prolonged from a stout circular band which surrounded its base.

Although spasm of the circular muscular fibres has been given above as one of the causes of intestinal stricture, and although it doubtless does form a very important element in many cases of fatal obstruction of the bowels, it is certainly of very rare occurrence, as an independent affection, and may be considered practically as limited to the rectum and anus. And indeed, even in these parts, spasmodic obstruction is probably always attended with some ulceration of the adjacent mucous membrane, to which there is reason to believe it secondary. Thus spasmodic contraction of the sphincter ani, an affection which may be regarded as exclusively surgical, seems to be dependent on the formation of an ulcer, at or within the verge of the anus; and not very infrequently spasmodic contraction, with great hypertrophy of the muscular tissue, is met with as one of the troublesome sequelæ of dysenteric ulceration of the rectum.

But the varieties of stricture with which we have here to deal particularly, are those in which, according to the definition with which we started, the stricture is due either to the contraction of the mucous and submucous tissues, or to some deposit or growth involving the general thickness of the walls. The cicatrization which follows ulcerative destruction of the mucous membrane is a common cause of diminution of the calibre of the bowel. But what particular kinds of ulceration are most apt to be followed by this condition is not very clear. Indeed, in most cases where stricture from ulceration is found after death, there is nothing in the history to guide our judgment in this respect. It is certain, however, that in order to produce any marked constriction, the area of ulceration must either have been considerable, or must have extended round the bowel. There is reason to believe that irritant poisons, in consequence of their corrosive effects on the mucous membrane, lead occasionally to the production of stricture of the intestine, especially in its upper part, just as they occasionally cause œsophageal stricture. There is no doubt that tubercular ulceration of the bowels, which very commonly forms annular patches or occupies extensive tracts, and which not at all infrequently undergoes more or less perfect cicatrization, is a yet more frequent cause of stricture, either in the lower part of the ileum, or in the cæcum, or in some part of the colon. Dysenteric ulceration of the large intestine is also a distinct cause of stricture; as again is the separation by sloughing of an invaginated portion of bowel. The ulcers of typhoid fever, on the other hand, are known to result very rarely, if ever, in obvious contraction of the calibre of the bowel: although it is pretty certain that even in this case, when the ulceration has spread and become extensive, marked constriction may attend its cicatrization. When stricture is due to ulceration, we find the mucous surface contracted, sometimes completely cicatrized, sometimes presenting unhealed spots of ulceration, with fungous excrescence or granulations, and separated from the subjacent muscular coat by a more or less abundant deposit of dense fibroid tissue. The stricture itself may be a mere ring, or it may occupy several inches of the length of the bowel; I have seen the whole cæcum thus reduced into a channel barely capable of admitting a goose's quill. Another cause of stricture, limited probably to the large intestine, is the growth of that fibroid material which resembles, but has of late been distinguished from, true scirrhus. This generally involves all the coats to a greater or less extent, encroaching, as it grows, upon the intestinal tube. Sometimes, but not necessarily, its surface ulcerates. A growth probably identical with this, occurring in so-called "pelvic cellulitis," sometimes involves the walls of the rectum and causes stricture there. But by far the most frequent cause of stricture is the development of cancerous disease in the coats of the intestine. This is sometimes local, or at all events of primary origin in the bowel, being then, perhaps without exception, a disease of the large intestine; but more frequently it involves the gut by spreading to it from some neighboring part, as from the peritoneum, the mesenteric or other abdominal lymphatic glands, from the substance of the gastro-hepatic omentum, from the cellular tissue of the venter ilei or pelvis, or from the genitourinary organs.

The presence of a stricture is always a more or less serious impediment to the progress of fæcal matters along the bowel; and in all cases therefore leads in a greater or less degree to certain results. These are : first, undue accumulation of fæcal matter above the stricture, with proportionate dilatation of the bowel there; second, hypertrophy of the muscular parietes of the dilated bowel; and third, diminution in calibre and even atrophy of the bowel below. It is an interesting fact that, in cases of stricture of the colon, the greatest degree of dilatation is often found, not in the portion of intestine immediately above the stricture, but in the cæcum. The tighter and the longer a stricture, the more exaggerated, other things being equal, will be the several consequences just described; and the more danger will there be of the supervention of permanent obstruction. Yet it is a very remarkable fact, that very tight strictures are not infrequently found after death in cases in which during life there has been no suspicion of their presence. Allusion has been already made to a case which was under my own care, wherein the cæcum was contracted into a channel two inches long, and about the size of a goose's quill; yet the patient had no symptoms of stricture, and died of acute pneumonia. But it is in the small intestine especially that stricture is apt to be present without producing any of its characteristic symptoms—a phenomenon which is probably due, in part, at least, to the fact that the contents of the small intestine are usually much more fluid than those of the large, and are consequently much more readily propelled through a very narrow orifice. Indeed, Dr. Buchanan's case already cited, and many others that might be quoted, show clearly, what also common sense would lead us to surmise, that the more solid the matters are which ought to be forced through a stricture, the more likely are they to be arrested there, and thus to render the obstruction complete. It may be added, that the lodgment of fæces above a stricture is very apt, not only to prevent the complete cicatrization of the ulcer by which the stricture itself may have been originally produced, but to cause erosion and ulceration in the dilated bowel above, a contingency which is still more likely to arise when cherry-stones or plum-stones or other hard bodies form a part of the accumulation. And, further, it may be added, that perforation of the bowel at or above the seat of stricture is not of very infrequent occurrence, generally as the result of perforating ulcer, occasionally as the result of laceration from associated softening and over-distention.

Stricture may be met with in any part of the intestine, yet it occurs in different parts with very different degrees of frequency. The published statistics of fatal cases show that its occurrence as a fatal disease in the small intestine is comparatively rare (according to Dr. Brinton,[1] in 8 out of every 100 cases) and that as regards the large intestine (to quote again Dr. Brinton's figures, with which those of other writers agree pretty closely), out of 100 fatal cases, 4 are in the cæcum, 10 in the ascending colon, 11 in the transverse colon, 14 in the descending colon, 30 in the sigmoid flexure, and 30 in the rectum. Dr. Brinton calculates that stricture occurs three times in men to twice in women; and that the average age of death is 44½ years.

(b) The Symptoms to which stricture gives rise vary greatly according to circumstances, especially according to its position, its degree, its cause, and its complications. As has been already pointed out, stricture of the small intestine very rarely causes symptoms sufficiently characteristic to enable us to diagnose its presence, and rarely causes death except by the accession of complications which themselves are not distinctive. It probably gives a liability to colicky pains, and to some degree of nausea and sickness. Indeed, in the case of the large intestine the symptoms

---

[1] "Intestinal Obstruction," by William Brinton, M.D., F.R.S. 1867. Frequent reference is made to this work throughout the present article.

produced by stricture may be for a long time vague and inconclusive, and even misleading. The patient suffers perhaps for weeks, or· months, or years, with occasional attacks of colicky pain, associated, it may be, with more or less constipation; but not infrequently during the earlier period of his malady diarrhœa may be a yet more prominent symptom. If, however, the obstruction be in the vicinity of the rectum, solid motions generally soon assume a narrow tape-like or pipe-like form. Occasionally the symptoms of obstruction come on quite suddenly; but most frequently some degree of constipation long precedes the occurrence of complete obstruction; and sometimes, too, it happens that the patient, previous to his final attack, may have experienced one or two or more similar atttacks, which have, however, yielded to treatment. The symptoms which attend and indicate impassable stricture are insuperable constipation, painful peris· talsis coming on periodically, and often rendering itself audible by borbo· rygmi, and visible through the abdominal walls, abdominal fulness and uneasiness, followed after a time by nausea and vomiting—the vomited matters becoming finally stercoraceous—and death at last from simple asthenia. Febrile symptoms and abdominal tenderness are often absent from first to last: but sometimes inflammation supervenes, or perforation takes place, and then enteritic or peritonitic symptoms become superadded. When the case is free from these or other complications, its progress is essentially chronic, and the patient, if not improperly treated, lives for a considerable time, often for many weeks. The duration of life in these cases may be said somewhat roughly to vary between two weeks and three months. Indeed, when we consider that constipation may continue for three months or more with comparatively little injury to the system, it is impossible not to believe that persons with simple impassable stricture of the rectum may, under favorable circumstances, survive for even a longer period than that.

It is always satisfactory, and sometimes highly important, to ascertain the exact site of stricture; and in coming to a conclusion on this point, it is well to bear in mind that at least three-fourths of the strictures of the large intestine are situated to the left of the mesial line of the abdomen. We need not, however, in all cases limit ourselves to a simple calculation of chances. It is natural to believe that the distention of the bowel above the stricture, and its collapse below, should reveal themselves to manual if not to ocular examination of the abdomen, and sometimes, no doubt, the form and position of a struggling, or even of a quiescent, length of distended bowel, may by such means be clearly identified. Fulness and dulness and weight in the course of the cæcum and ascending colon, or on the right side of the belly, might thus indicate a stricture at or about the hepatic flexure, and, associated with the same conditions extending across the epigastrium, might indicate stricture at the splenic flexure or in the descending colon; whereas fulness, and the like, limited to the left side of the belly, or most pronounced in that region, might equally be indicative of stricture in the sigmoid flexure or rectum. But thickness or rigidity of the abdominal walls, or tenderness, or the presence of tumors, or the altered positions which greatly distended tracts of bowel are apt to assume, often interfere to prevent the easy recognition of even extreme differences of intestinal dilatation and fulness. Dr. Brinton maintains that the amount of fluid which may with care be injected *per anum*, is a very valuable guide in estimating the point of stricture. He says: "With a maximum injection of a pint of warm bland liquid, the obstruction of an ordinary male adult may be referred to a point not lower than the upper end of the rectum. A pint and a half, two pints, three pints, belong to corresponding segments of the sigmoid flexure. The descending and transverse colon accept a larger but more irregular quantity." But here again there is evidently very abundant room for error; for it is certain that not all contracted bowels are tolerant in an equal degree of mechanical distention, and there can be no doubt! that a stricture, which may prevent the passage of hard fæcal matter in one direction, may yet allow of the transmission of thin fluids in the opposite direction. Lastly, when the stricture is a short distance only from the anus, its presence may often be ascertained by the introduction of the finger, or, as has been suggested, of the entire hand; and if it be beyond the reach of actual touch, yet in the rectum, the careful introduction of a bougie may perhaps reveal its position. But it must not be forgotten that the curvatures of the rectum, and the prominent folds of its mucous membrane, are such impediments to this latter mode of examination as to rob it of very much of its value; in addition to which, it is attended with, at all events in many cases, considerable risk of damage.

(c) *Treatment.*—Whenever we have reason to believe in the presence of a stricture, it is obviously desirable that nothing which is not in a perfectly fluid or pultaceous condition should be allowed to enter the bowel, —therefore, that the food taken habitually should be easy of digestion, thoroughly well masticated, and not more abundant than is absolutely necessary for the preservation of health, and especially that neither plum nor cherry stones, nor even pips, should be swallowed: secondly, that the bowels should themselves be kept as far as possible in a quiet condition—in other words, whilst constipation should as far as possible be prevented, diarrhœa and painful gripings should equally be guarded against. If there be constipation, it may be directly relieved, or the bowel above the seat of stricture may be encouraged, as it were, to propel its contents by the use of simple non-purgative enemata; but purgatives of all kinds, certainly anything like active purgation, should be religiously eschewed. Should the stricture be in the rectum, and within reach, it may of course admit of dilatation and relief by the use of a bougie. When symptoms indicative of complete stoppage manifest themselves, the wish to employ active measures to relieve the patient's distress naturally obtrudes itself; but such measures are for the most part even less admissible now than formerly. Enemata may be of advantage, partly, as before pointed out, to guide our judgment as to the seat of the stricture, partly (if the stricture be in the large intestine) for the purpose of promoting the relief of the bowel above the impediment; but purgatives are not only useless, but almost certain to do serious mischief, if not to cause actual perforation. On opium and other sedatives, and soothing applications locally applied, utterly inadequate though they generally are, must yet be our chief reliance, so far as ordinary medical treatment is concerned. But in all such cases a time comes when the advisability of forming a communication from without with the portion of bowel above the stricture—in other words, the attempt to establish an artificial anus—becomes a serious question. When the stricture is in the large intestine, as it generally is, Amussat's operation, in one or other loin, is that which would of course be chosen for performance; and although it is obviously incompetent to cure the stricture, it avails very often to prolong life, and sometimes to prolong it for a considerable period. If the stricture happens to be in the small in-, testine, Litré's operation is alone available.

III. COMPRESSION AND TRACTION.—Dr. Hilton Fagge[1] has with great reason distinguished on the one hand from stricture, on the other from internal strangulation, a class of cases related to both, which is yet clearly distinguishable from them, and which he designates "Contractions." They are cases in which the bowel becomes obstructed by the compression, or the pressure, or the traction exerted upon it by adhesions, or growths, or deposits situated externally to it, and in which there is no contraction inherent in the walls themselves, and not necessarily or generally any strangulation.

(a) *Pathology.*—Under the above heading may be included those cases in which the return becomes obstructed, and defæcation rendered painful or difficult, in consequence of the pressure exerted on that part of the bowel, either by an enlarged or displaced uterus, or by a uterine or ovarian tumor. It is conceivable, of course, that any form of abdominal tumor may by pressure obstruct the alimentary canal in some part of its course. I recollect one case of death by rupture of the abdominal aorta, in which the blood, effused and coagulated in the sub-peritoneal tissue, had so surrounded and compressed the third part of the duodenum that the finger passed along it with difficulty; and while the stomach and duodenum above contained a considerable quantity of contents, the intestine below was perfectly empty.

But the cases which are now more particularly referred to are those in which obstruction is due to the embarrassment of a greater or less length of bowel, caused by the presence on its outer surface of lymph or false membrane, which binds it more or less firmly to the surrounding parts, and sometimes constricts, sometimes leads to the formation of sharp angular bends. The adhesions are often produced by circumscribed peritonitis, but more frequently, perhaps, are developed in the course of peritoneal tubercle or cancer. In some cases the intestine has been incarcerated in a hernia, and portions of it have become invested in adhesions, which attach it, perhaps, to the neck or some other part of the sac, or to the omentum; in others, the transverse colon or sigmoid flexure, or some other tract of bowel, is hooked down, as it were, by bands of lymph to the uterus, or ovary, or some other structure within the pelvis; in others, again, several contiguous coils of small intestine are tightly bound together, forming a kind of tangled mass. Fatal cases of compression or traction always furnish distinct evidence of more or less complete obstruction, in the contraction and emptiness of the bowel below, and in the dilatation, hypertrophy, and fulness of the bowel above; but the part in which the actual obstruction has taken place, though contracted and more or less empty, is frequently found to admit with ease of the passage of the finger, or even of some larger body. The immediate cause of obstruction indeed is not generally a simple tight constriction, but consists sometimes in a comparatively slight compression of a considerable length of bowel, which thus becomes embarrassed in its action, and sometimes in the presence of a sudden bend or twist, the upper portion of which becoming distended presses upon and flattens the portion beyond, and so renders it impervious, and in association with these doubtless a greater or less degree of spasmodic contraction. Sometimes, however, the obstruction is as sharp and definite as any stricture.

Dr. Fagge points out (and in the opinion which he expresses I entirely

---

[1] In an excellent paper in the Guy's Hospital Reports for 1869, to which frequent reference is made in the course of this article.

agree with him) that these cases are of far more frequent occurrence in the small intestine than in the large, and that in a clinical point of view they may be regarded as the strictures of the smaller bowel.

(b) *Symptoms and Treatment.*—The symptoms of the affection now under consideration are almost, if not quite, identical with those of stricture. In both cases, when the impediment to the due action of the bowel is associated with abdominal cancer or tubercle, or any other form of adventitious growth, the symptoms connected with these complications mask, if they do not conceal, the symptoms due to obstruction. In both cases, when no such complications are present, the symptoms sometimes come on quite suddenly, sometimes creep on insidiously with occasional colicky pains, limited but powerful peristaltic movements, and gradually increasing obstinacy of the bowels; and sometimes the patient suffers from one or more severe attacks of total constipation, which yield after a time to nature or to treatment, and in this respect only differ from the final and fatal attack. In both cases, again, the disease, though not entirely free from the danger of the supervention of peritonitis or enteritis, is still not necessarily complicated with symptoms of inflammation, and its course, therefore, tends to be peculiarly chronic, lasting sometimes for weeks, and its close is usually determined by gradual exhaustion only. Dr. Fagge thinks that cases of this kind are to be distinguished by their chronicity, by the occurrence of obstruction rather in the small intestine than in the large, and by the powerful and well-marked vermicular movements which occur, often nearly to the last, in the length of bowel above the impediment. He points out that it is in cases of chronic impediment especially that the bowel above becomes hypertrophied as well as dilated, and he argues that it is therefore probably in these same cases (stricture and compression) that the movements of the bowel, in their endeavors to overcome the impediment, are most powerful and most obvious. In confirmation of this view, I may state that the cases in which I have myself most distinctly traced the peristaltic movement of the bowel have been cases of the kind in question.

It is needless to draw any distinction here as regards treatment between stricture and compression of the bowel.

The following case may be quoted as a typical example of the affection which has just been described. A man, forty years of age, was attacked suddenly, seven weeks before his admission into St. Thomas's, with severe colicky pains, which confined him to his bed for two or three days. He improved, but at the end of a few days had a recurrence of the same symptoms, lasting for about three weeks, and attended with nausea, vomiting, and constipation. Then for ten days he became free from pain and apparently convalescent. But ten days before his admission all his symptoms returned with increased severity; and during this time vomiting was pretty constant and his bowels remained unopened, although strong purgatives were several times administered. On admission his face was anxious, but his tongue was clean and his pulse quiet. He vomited regularly two hours after taking food. The belly was distended and tympanitic, and somewhat tender; he complained of constant pain in it; and severe exacerbations of pain, lasting two or three minutes, and attended with a gurgling sound, came on about every five minutes. The vomiting became stercoraceous four days after admission, and continued so thenceforth. The bowels were never acted on except by enemata, which brought away fæcal matters in gradually decreasing quantities. The distention and tenderness of the belly continued, if they did not increase; and the paroxysms of more intense pain coming on every few minutes troubled him almost to the last. During these paroxysms, the violent peristaltic movements of the bowels could be followed through the abdominal parietes with the greatest facility. He had no distinct febrile symptoms, and no hiccough; he continued perfectly sensible, and died of simple exhaustion just three weeks after admission. At the post-mortem examination, the small intestines generally were found to be enormously distended, and their surface a little heightened in color, and marked with longitudinal bands of rather intense capillary congestion. From the middle of the ileum to within a foot of the cæcum the coils were adherent to one another and to the brim of the pelvis by bands and filaments of false membrane, and were so entangled that their direction was traceable with difficulty. The portion of bowel involved was for the most part somewhat dilated; its lowest third, however, was contracted and empty, as also was the portion between this and the cæcum. The stomach and small intestines down to the seat of contraction were dilated, and full of thin pea soup-like fluid; the cæcum and large intestines were contracted throughout, but here and there in the ascending colon were small lumps of hardened fæces. The mucous membrane of the alimentary canal was healthy everywhere. There was no hernia, no intussusception, and no part of the bowel along which the finger could not readily be passed.

IV. INTERNAL STRANGULATION.—Internal Strangulation arises from similar causes to those which produce ordinary strangulated hernia, namely, constriction or nipping of a portion of bowel by the edges of some natural or artificial orifice through which it protrudes, with consequent arrest of the circulation of blood in it, and impediment to the passage of fæcal matters along it. Such orifices are the foramen of Winslow, congenital or acquired perforations in the mesentery, meso-colon, great omentum, or other peritoneal duplicatures, or apertures formed, with the aid of neighboring parts, by bands of fibroid tissue (the result generally of some inflammatory process) extending from one point of the peritoneal surface to another. And it is obvious that the same accidental conditions which lead to the protrusion of intestine into an ordinary hernial sac, may equally lead to the protrusion of a knuckle or loop or still larger mass of bowels into one of these. But, of course, it no more follows in the one case than in the other that strangulation should either immediately, or at any subsequent period, follow upon this displacement; although in both cases there is always imminent danger of its occurrence.

(a) *Pathology.*—Protrusion of bowels through the foramen of Winslow must be an exceedingly rare event. Rokitansky,[1] however, alludes to a case in which he found this the cause of strangulation of a large portion of small intestine. Perforation of the various duplicatures of peritoneum, with the passage of intestine through the perforation, and consequent strangulation, is a much more frequent occurrence. This accident appears to be most common in connection with the mesentery, and then generally to follow upon laceration from violence. Next probably in order of frequency it is met with in connection with the great omentum. And cases are recorded in which death has followed the strangulation of a portion of bowel through a hole in the duplicature of peritoneum belonging to the vermiform appendix, or through a hole in the suspensory ligament of the liver, or in the broad ligament of the uterus. Meso-colic rupture is probably a congenital malformation. Three cases of it are recorded in

---

[1] Pathological Anatomy : Sydenham Society's Translation, vol. ii.

the "Transactions of the Pathological Society;" and in each of them nearly the whole mass of small intestines was contained in a large pouch of the transverse meso-colon, or in the mesentery of the transverse and descending colon. In two of them death was due to disease independent of the rupture; in the third, recorded by Dr. Peacock, the patient died of strangulation. There is probably no part of the peritoneal surface to which bands capable of producing strangulation may not be attached; but there are certain structures and certain conditions of parts with which they are specially apt to be connected. Thus, the vermiform appendix often becomes adherent to neighboring structures, such as the mesentery, small intestine, colon and ovary, forming a kind of loop; thus, too, diverticula of the lower extremity of the ileum become attached, with a similar result, usually by the apex, either to the mesentery or some other neighboring part, or are prolonged to the umbilicus in the form of a cord (a remnant of fœtal life). Again, bands producing strangulation are often joined to the mesentery, or the parts concerned in old ruptures, and often to the pelvic organs, more particularly the uterus, Fallopian tubes, and ovaries. It may here be noted also that strangulation is not very infrequently produced by the slipping of a loop of intestine under the lower edge of the mesentery (unusually elongated), of a portion of bowel hanging low into the pelvis, or even under the pedicle of an ovarian or uterine tumor. Finally, there are rare cases of internal strangulation, in which the bowel protrudes into a lacerated bladder or uterus, or into a perforated bowel, or through the diaphragm. Cases also are occasionally met with in which there is a free communication, generally, if not always congenital, between the peritoneum and pericardium, or one of the pleuræ.

The small intestine is much more frequently strangulated than the large; and of the large intestine the regions most liable to this accident are those which are most movable, namely, the cæcum and sigmoid flexure. Internal strangulation occurs at any age; generally, however, above thirty; but strangulation in connection with the appendix vermiformis, or a diverticulum happens most frequently in comparatively early life, the average age being, according to Dr. Brinton, twenty-two years; further, strangulation from diverticula and from lacerated mesentery is, according to all authorities, far more common among males than females. It has already been pointed out that there is a very important relation between peritoneal bands and the sacs of old herniæ, and in females between such bands and the pelvic organs.

(b) *Symptoms.*—The symptoms of internal strangulation are identical with those of ordinary strangulated hernia, and so like those which have been described as the symptoms of the severer form of enteritis that there is no occasion to give here any special account of them. It need scarcely be added that they differ essentially from those of stricture and of compression of the bowel, in the facts that they are always sudden in their origin and acute in their severity and progress, and always end fatally (if the stricture be not relieved) within a few, rarely more than five or six, days.

(c) *Treatment.* — As regards the general management and medical treatment of these cases, nothing can be added to what has already been laid down in reference to enteritis. But whenever the diagnosis of an internal strangulation has been made, it must of necessity become a question whether an operation should be performed with the object of relieving it. There can be no doubt, of course, that the liberation of a portion of bowel strangulated by any of the various causes above enumerated

ought *cæteris paribus* to be attended with as good results as the division
of the stricture in ordinary cases of strangulated hernia; but there is also
no doubt that operations performed with that intention have not on the
whole afforded any encouraging results. When, however, we consider
that although typical cases of the different kinds of intestinal obstruction
may really present characteristic peculiarities of symptoms, it is yet for
the most part exceedingly difficult in practice to discriminate the cases
that come before us, and that therefore operations must comparatively
often be performed where from the nature of things they must be useless;
and, further, that while even in the case of the operation for ordinary
strangulated hernia its early performance is generally essential for its
success, in the case of internal strangulation the operation, if performed
at all, is almost always delayed until a late stage in the disease; it is not
hard to understand why so little success has attended the operative treatment of the cases under consideration. A sufficient number of operations
has, however, been successful to justify us in laying it down as a rule,
first, that in every case in which we have come to the conclusion that a
patient is suffering from internal strangulation, an operation should be
performed for its relief; secondly, that in all cases in which we think it
not improbable that such a strangulation exists, the patient should not be
allowed to die without an exploratory operation having been effected or
at least proposed.

(*d*) *Note on Torsion or Twisting of Bowel.*—There is a class of cases,
far from uncommon, which may be conveniently adverted to here. They
are cases of what is called "Torsion" or "Twisting" of the bowel. It
has already been shown that fatal obstruction to the passage of fæcal
matters along the bowel may be caused, or appear to be caused, by the
formation of some abnormal abrupt bend, or twist, in connection usually
with external adhesions. In these cases, however, death is caused, as in
stricture or compression from without (with which last I have classed
them), by obstruction alone. But in the cases now to be considered, the
twisted portion of bowel becomes the seat of enteritis, and death results
speedily, with the symptoms of enteritis rather than those of obstruction.
The cases, indeed, clinically seem to be undistinguishable from cases of
enteritis or internal strangulation. The onset of the disease is sudden,
the symptoms acute and severe, and the supervention of collapse and
death speedy. And on examination after death there is found a length
of bowel greatly dilated and black with congestion and inflammation, if
not gangrene, no strangulation, at least no strangulation in the ordinary
sense of the word, but instead, a remarkable twisting of the inflamed
tract of bowel with its mesentery, by which twisting it is supposed that
the vessels leading to and from the part have become obstructed. Such
twisting, associated with inflammatory mischief, is sometimes observed in
the small intestine; but it is far more commonly met with in connection
with the larger bowel, and especially with the sigmoid flexure and cæcum.
If these cases be really, as is generally believed, cases in which strangulation of the bowel is produced by the twisting of itself and its mesentery,
they naturally fall under the head of internal strangulation, with which,
as has been pointed out, their symptoms and progress ally them. I must
confess, however, that I have a strong inclination to believe that most, if
not all, recorded cases of this affection are essentially cases of enteritis,
and that the twisting is a secondary phenomenon only. It is not very
easy to see how a portion of bowel, unless its position be altered and its
movements interfered with by adhesions (and certainly in many of the

cases no adhesions whatever are observed), can become so twisted by any movements of its own, or even by the pressure of surrounding healthy parts, as to be either strangulated or incapable in virtue of its own peristaltic movements of recovering its normal position; but it is easy to see that an inflamed and paralyzed portion of intestine, heavy with accumulated contents, dilated to many times its normal bulk, and forming a doughy, inelastic, inert mass, may under certain conditions by its mere weight subside from its normal site, or be pushed aside by the pressure of the actively vital parts around it, and so be made to assume a position and form suggesting the generally received explanation of the sequence of events.

V. IMPACTION OF FOREIGN BODIES.—It has already been pointed out that mere ordinary intestinal contents, no matter how unwholesome, how indigestible, or how imperfectly comminuted the ingesta from which they are derived may be, very rarely indeed cause by their accumulation permanent intestinal obstruction; yet it is not improbable that, according to the ordinary belief, undigested masses of food do sometimes, in their passage along the small intestine, move with difficulty, or become temporarily impacted, and so produce pain and sickness, and even symptoms of obstruction. Dr. Brinton describes a case of this kind, in which he asserts that he distinctly traced by palpation a mass of half-chewed filberts in its passage (lasting two days) along the small intestine.

(a) *Pathology.*—Foreign bodies, indeed, of comparatively small size, such as coins, fragments of bone, teeth, marbles, plum-stones and cherry-stones, generally pass along the healthy intestine without causing any material inconvenience; and occasionally even pointed bodies—pins and the like—prove equally innocuous. They are all, however, a source of serious danger in the presence of strictures, above which they usually become arrested, or in which they may become lodged. The smaller ones among them may lead also to serious results by slipping into a diverticulum, or into the vermiform appendix; and those which are pointed are apt to perforate the intestinal wall, and thus, escaping into the peritoneal cavity, to set up fatal peritonitis, or, escaping into the surrounding tissues, to provoke suppuration there. In the latter case, the foreign body sometimes emerges through the abdominal parietes, sometimes (when it perforates the rectum) is the cause of anal fistula.

Insoluble matters, in the form of powder, or in a fibrous state, which under ordinary conditions may be swallowed with perfect impunity, occasionally, after having been taken habitually in large quantities and for long periods, are found to have been gradually deposited from the fæcal contents of the bowels, and to have concreted into hard masses. These are sometimes round or ovoid, and may then be termed intestinal calculi, and sometimes form casts of the portion of gut in which they lie. The former are probably always found in the large intestine: the latter rarely, if ever, occupy any other position than the rectum. Among substances which thus occasionally form concretions, are sesquioxide of iron, carbonate of magnesia, insufficiently cooked starch, and oat-hair derived from oat-cake and other articles of food made from oats.

Amongst cases of exceptional rarity may be included those which are here and there recorded of persons who have been in the habit of swallowing knives, or pins, or string, or hair, or cocoa-nut fibres; things which, from various causes, are somewhat difficult of transmission, and which with the constant additions which are made to them gradually form accumulations or masses, which sometimes attain very considerable dimen-

3

sions, and may then easily be distinguished through the abdominal walls. These are generally found to occupy more particularly the stomach and upper part of the small intestine, and, when composed of fibrous substances, take the shape of the cavity in which they have formed. Their presence causes gradual dilatation of the part in which they are lodged, then congestion, inflammation, and ulceration, and finally, either perforation into the peritoneal cavity, or complete obstruction. It is remarkable, however, how long a period often elapses before such cases terminate in death, and how little, comparatively, of distress or even inconvenience the patient often experiences previous to the supervention of fatal symptoms.

But the usual cause of fatal impaction, and that which comes more especially within the scope of the present article, is the escape of a large gall-stone from the gall-bladder into the small intestine. The gall-stones here referred to are not those which so commonly slip from the gall-bladder into the cystic duct, and thence into the common duct, and thence (if they do not become firmly fixed there) into the duodenum ; for although these cause grave symptoms enough so long as they are retained within the biliary passages, they cease, as a rule, to cause any ill effects so soon as they have gained an entrance into the bowel ; their comparative smallness allowing them to pass along the intestines and to escape with the fæces, just as a plum-stone or a cherry-stone might do. The biliary concretions which become impacted in the bowel are single stones, or masses of coherent stones, of considerable bulk, varying, at a rough estimate, from three to four inches in circumference, and from one inch to two, three, or even four inches in length; in the former case presenting more or less of the ordinary cuboidal form, in the latter case forming a more or less complete cast of the gall-bladder. It is obviously scarcely possible that concretions of this magnitude can escape from the gall-bladder *per vias naturales;* and there is reason to believe that in all cases where a careful examination has been made, an ulcerated opening has been discovered, by which the cavity of the gall-bladder and that of the duodenum were in tolerably free communication, and through which the concretion had obviously escaped from its bed. When a large calculus has thus got into the duodenum, it seems to be carried on with the other contents of the bowel by means of the ordinary peristaltic movements. But its mere bulk prevents it from moving readily: besides which it provokes by its shape and hardness, as well as by its bulk, some irritation, if not inflammation, of the mucous surface over which it passes, and more or less spasmodic contraction of the muscular tissue which surrounds it. It hence continues to progress irregularly, now moving slowly, now coming to a standstill, impelled onwards by the *vis à tergo*, checked in its passage by the spasmodic contraction of the portion of bowel which embraces it, and by the comparatively empty and contracted state of that which is below it, and causing as it descends more and more mischief to the mucous surface, until finally it becomes impacted, sometimes in the jejunum, sometimes in the ileum, and not unfrequently just above the ileocæcal valve. Then all the effects of complete obstruction, conjoined with those of intense enteritis, supervene ; the bowel below becomes empty, that above distended with accumulated contents, and generally more or less inflamed, while at the seat of obstruction and in its immediate neighborhood the inflammation becomes intense, extending speedily to the peritoneal surface, and ends not rarely in gangrene and in perforation. Gallstones rarely, if ever, become lodged in the cæcum, colon, or any other part of the large intestine.

Gall-stones are a product of the later period of life ; and hence obstruction by gall-stones can only be looked for at an advanced age. It occurs indeed rarely before the fiftieth year, and, it may be added, much more frequently in women than in men. Dr. Brinton estimates the average age of its occurrence at 53⅓ years, and that it occurs four times as frequently in women as in the opposite sex.

(b) *Symptoms and Treatment.*—The symptoms which indicate obstruction of the bowels by a gall-stone are as nearly as possible identical with those which attend internal strangulation or enteritis. The cases themselves are, however, amongst the most violent in their symptoms and the most rapid in their course of all cases of intestinal obstruction ; conditions which result partly from the intensity of the inflammation which attends them, partly from the fact that the obstruction is almost without exception situated in the small intestine, and often high up in it. Dr. Brinton calculates their average duration at five days. There are two or three circumstances which may afford more or less assistance in the discrimination of obstructions by gall-stones : such are, first, the age and sex of the person attacked; second, the possibility in certain cases of discovering by palpation the presence of a gall-stone (that is to say, of a solid mass) in the bowels, and even of tracing in some degree its progress; and third, the occurrence of precursory symptoms due to the escape of the gall-stone from the gall-bladder, and to its presence in the bowel in the interval between this escape and its final impaction. It must not be forgotten, however, that in practice not only do we often fail in these cases to recognize a lump, or to obtain a history of premonitory symptoms; but that we may have both a lump and a history in cases where the symptoms are wholly independent of the presence of a biliary calculus or other foreign body. There does not appear to have been observed any connection between ordinary " attacks of gall-stones," and the affection now under consideration. This circumstance, however, is not remarkable, when it is borne in mind that gall-stones which escape by the normal route must necessarily be small, and that the escape of one such stone makes the way of escape for others that may be in the bladder comparatively easy, whereas those which cause intestinal obstructions are always large, and are often casts of the gall-bladder.

It may be added here, that not all large gall-stones cause death, after their entrance into the bowel, by obstructing it. They sometimes become encysted in a pouch which they have themselves been instrumental in producing. Dr. George Harley[1] records a case in which a gall-stone became thus lodged in the duodenum. Sometimes, again, they escape *per anum.* It is of course impossible to lay down any law as to the limits of size beyond which it is impossible for a solid body to pass through the ileo-cæcal orifice; but there are good grounds to suspect that in most cases where large calculi have been voided, they have passed by ulceration directly from the gall-bladder into the colon.

No distinction need be made between the treatment of cases of obstruction by gall-stones, and that of cases of enteritis.

VI.—INTUSSUSCEPTION.—(a) *Pathology.*—By intussusception is meant the prolapse or slipping of a tuck of intestine into the cavity of the portion of intestinal tube immediately below it, wherewith it is continuous. In consequence of this, we find the normal course of the intestine interrupted by a kind of knot, in which three successive lengths of bowel lie

---

almost concentrically one within the other; the innermost length being formed by the portion of bowel which has descended, the outermost length consisting of the portion of bowel into which the descent has occurred, the middle or intermediate length being the portion of bowel which unites the upper extremity of the one with the lower extremity of the other, and lies therefore in an inverted and everted position between them. The mesentery of the inner two, or included, lengths of bowel is in their descent necessarily dragged down with them into the pouch which they form, and, by the unilateral traction which it exerts, necessarily gives to their double tube a curvature of which the concavity corresponds to the line of mesenteric attachment; so that the lower orifice of the invaginated portion of bowel, instead of lying in the axis of the containing bowel, faces and rests upon some portion of its circumference. The several layers are generally more or less convoluted (with convolutions running transversely) or twisted: but this convolution or twisting is always most marked in the middle tube. The immediate effects of intussusception are, first, more or less obstruction to the passage of the intestinal contents, and, second, more or less obstruction to the return of blood from the inner two cylinders of bowel involved, to which the stretched and constricted portion of mesentery belongs. It is obvious that the innermost tube must be pretty tightly compressed by the tubes external to it, a condition which must be much increased by the swelling of parts which speedily takes place; especially it is always found to be very tightly girded at its point of entrance by the tumid ring formed at the junction of the outer two layers. Nevertheless, the obstacle which an intussusception opposes is often incomplete, for it is certain that in a good many cases fæcal matters, not always in small quantities, pass through it pretty constantly: a circumstance due, in part, to the efficiency of the contractile force of the bowel above to squeeze a portion of its contents into the narrowed tube below, but chiefly to the retention still of contractile power in the affected portions of bowel. Very soon after the occurrence of intussusception all the tissues of the inner two tubes, internal to the serous membrane, become black or nearly so with congestion and escape of blood into their substance, and the serous surface consequently assumes a more or less deep slate-color. At the same time, partly from the accumulation of blood, partly from the transudation of serum, their walls become very greatly swollen, and sanguinolent serum or blood becomes effused from the mucous membrane, and may be found collected both in the interval between the opposed mucous surfaces of the outer two layers of the intussusception, in the central canal, and in the bowel below the seat of disease. At a somewhat later period coagulable lymph is secreted from the opposed serous surfaces of the middle and internal layers, and these become consequently agglutinated in their whole length. The two invaginated tubes remain sometimes for a long while in the condition above described, but often ere long become gangrenous, and then, if the patient survive sufficiently long, separate from their attachments and become discharged *per anum*.

Intussusception is doubtless always an accident of sudden occurrence in connection with some violent spasmodic action of the portion of bowel which becomes prolapsed. It seems certain, however, that there must be some associated conditions which concur with spasmodic action in producing it. A wave of peristalsis is made up of two distinct elements: first, the contraction of the longitudinal fibres which shortens the bowel and dilates it, and (since it travels from above downwards) draws the portion of bowel below, in which the contraction is commencing, towards the portion of bowel above, in which the contraction is completed; second, immediately following upon this, the contraction of the circular fibres which narrows the bowel and elongates it, and, in elongating it, projects the narrowing segment forward. Now, it is obvious that in these two associated elements, namely, the dilatation of one segment of the bowel with a tendency in its lower part to be drawn upwards, and the narrowing of the segment of bowel immediately above it with a tendency in its lower end to be pushed forwards, we have conditions which, with very slight modification or exaggeration, might permit of the protrusion of the narrowing segment above into the dilated segment below. The circumstances which either alone or in combination might have this effect would seem to be: first, the presence of much gaseous matter leading momentarily to excessive distention of the portion of bowel into which the wave of circular contraction is advancing; second, immobility from whatever cause of this distended portion of bowel, so that it is not pushed on bodily by the elongation of the narrowing segment above; and third, the occurrence at this moment of some violent muscular effort, involving the action of the muscular parietes of the abdomen. The efficiency of these, or of equivalent circumstances, in causing descent of the bowel, is shown in the cases of prolapse of the rectum, and prolapse of bowel through an artificial anus; as well as in the most common case of intussusception, namely, that in which the extremity of the ileum slips into the cavity of the cæcum. It is supposed that the presence of lumbrici occasionally determines the occurrence of intussusception, and with more reason that the presence of a large polypus has this effect. It may be remarked, however, that in some of the recorded examples of concurrence of intussusception and polypus, the intussusception and polypus have been at a distance from one another.

In every case an intussusception must obviously in the first instance involve a short length of bowel only; but for the most part it rapidly increases in size owing to the active peristaltic movements of the several segments engaged. This increase takes place partly by the prolapse of more and more bowel from above, but chiefly by the involution of more and more of the outer layer. In most cases indeed, if not in all, the parts which in the first instance formed the margins of the lower orifice of the invaginated portion of bowel continue to form that orifice, no matter what length the intussusception may ultimately attain. The growth of the innermost tube therefore is the result simply of the descent of more and more bowel from above, while the growth of the middle tube takes place at the expense of the outermost tube only, in consequence of its gradual inversion.

The length of bowel involved in an intussusception varies within wide limits. Including in our measurement the inner two layers only, or those which constitute the intussuscepted portion, the length varies from two to three inches up to three or four feet. A case indeed is quoted by Dr. Peacock,[1] in which, judging from the combined lengths of portions which escaped from time to time *per anum*, there is reason to believe the invagination had comprised twelve feet of bowel.

Intussusception is rather more than twice as common in males as in females, both before and after puberty. It occurs at all periods of life,

---

but is singular, amongst obstructive diseases, in the frequency with which it affects young children.

Intussusception is not very infrequently met with after death in persons (children and adults) in whom during life there had been no reason to suspect its presence, who have had no symptoms which can be attributed to it, and who have died of some totally different disease. In these cases the intussusceptions are always found in the small intestine—sometimes, indeed, two or three are met with in the same case—they are generally not above an inch or two long, are easy of reduction, and present little or no œdema or congestion. It is not impossible, as has often been suggested, that similar slight intussusceptions take place occasionally during good health, and having caused symptoms of more or less severity, undergo spontaneous evolution with restoration of the integrity of the bowel. Intussusceptions which prove fatal may occur in almost any part of the intestinal canal, but they occur in different regions, with very different degrees of frequency. Out of 100 fatal cases (according to Dr. Brinton's figures), 4 are jejunal, 28 iliac, 56 ileo-cæcal (that is, involving the cæcum together with the ileum and colon), and 12 colic, or originating in and involving the colon only. It must be noticed, however, that recoveries with separation of the intussuscepted bowel are much more numerous in those cases in which intussusception occurs in the small intestine than in those cases in which it involves the colon, a fact which renders it more than probable that the jejunal and iliac varieties form a larger proportion of the whole number of cases of intussusception than Dr. Brinton's figures might lead us to believe. It may be added, moreover, that intussusception occasionally begins in the rectum, of which Dr. H. Fagge quotes an example; and that prolapsus of the rectum, which in some cases involves the descent of the muscular wall together with that of the mucous membrane, is under these latter circumstances a true intussusception.

Jejunal or iliac intussusception is met with generally, if not exclusively, in adults. The average age of its occurrence is, according to Dr. Brinton, 34·6 years. It is here that the peculiar curvature of the invaginated part, due to the traction of the mesentery, is most observable; and it is here, owing probably to the comparative narrowness of the tube into which the invaginated portion of bowel descends, that strangulation and congestion are most speedy and most intense, and that sloughing and separation of the strangulated part are consequently most frequent. The length of bowel engaged in this form of invagination, although it may be as much as several feet, is generally less than in intussusceptions involving the large intestine.

Ileo-cæcal invagination occurs largely amongst young children, including babes of a few months old. Dr. Brinton considers that half the total number of cases are in children under seven years of age; and that the mean age of those affected by it is 18·57 years. It begins with the descent into the cavity of the cæcum of the lips of the ileo-cæcal orifice, which form henceforth the lower extremity of the invagination. As this increases, the descending ileo-cæcal orifice drags down with it more and more of the ileum to form the central tube, and inverts first the cæcum, and then a gradually increasing quantity of the colon, to form the inverted or middle layer; and still descending, finally in some cases reaches the rectum or even protrudes from the anus. It may be added that the orifice of the vermiform process necessarily retains its position relatively to the ileo-cæcal orifice, and that the process itself therefore lies at the bottom of the pouch between the inner and middle tubes. In ileo-cæcal invagination, which is that in which the greatest length of bowel may be engaged, there is generally much transverse folding of the several layers of intestine which form it, especially of the middle layer, which is also often much convoluted or twisted. Strangulation is comparatively much more rare here than in intussusception limited to the small intestine, doubtless because of the comparative roominess of the colon: and in a proportionate degree sloughing and discharge of the invaginated tissues are necessarily uncommon.

A variety of ileo-cæcal invagination of very rare occurrence is that in which the lower extremity of the small intestine descends into the cæcum through the ileo-cæcal orifice; the lips of the orifice not necessarily descending with it. Strangulation in this case is said to be generally sudden and complete, in consequence of the tightness with which the prolapsed bowel is gripped by the valve. Colic and rectal intussusceptions are comparatively infrequent, and differ little, except in the parts involved, from the ileo-cæcal form of the affection.

If the patient survive sufficiently long after the formation of an intussusception, events take place in connection with it which have already been briefly indicated. The peritoneal inflammation which by its products unites the opposed serous surfaces of the inner two layers, may spread beyond its primary seat, and cause more or less general peritonitis. Or, after these two lawers have become united, a further descent of bowel may take place, producing what is called a double intussusception—an intussusception, that is to say, in which the bowel above has slipped in the form of a second invagination into the canal of the primary invagination. Or, again, as Dr. Aitken[1] shows, the extremity of the curved invaginated portion of bowel may, by the constant pressure which it exerts against the side of the containing tube, cause at the seat of pressure ulceration and perforation of the intestinal wall. But by far the most interesting and important event is the sloughing and separation of the included layers of bowel. It has been shown that almost immediately after the occurrence of invagination, these become œdematous, intensely congested, and infiltrated with blood; and it might be supposed from the obstruction to which the vessels supplying them are exposed, that their death must necessarily speedily ensue. In many cases, however, patients live for weeks, and even months, after the occurrence of invagination, with no further changes in the contained tubes than those due to mere congestion and swelling, and die ultimately from the effects of invagination, the bowel never, even to the last, showing signs of either ulceration or gangrene. This (as has been stated) happens rarely, if ever, in intussusception limited to the small intestine, but it is very common in the case of ileo-cæcal and colic invagination. But in many instances, and (as has also been stated) far more frequently in the case of the small intestine than in that of the large, the deep congestion ends in the death of the intussuscepted portion; which then after a while, if the patient still survives, becomes detached either bit by bit or in mass, and gradually working its way downwards becomes expelled. This separation generally leaves the upper extremity of the outer tube of bowel firmly united, at the neck of the intussusception, with the lower extremity of the healthy bowel above, the line of union between the two being indicated by an annular fissure externally, and by a ring of ulceration on the mucous aspect, attended with more or less diminution of the calibre of the intestine, and to which sometimes por-

---

tions of the intussuscepted bowel still living and forming a sort of excrescence remain adherent. Sometimes at the moment of separation of the sequestrum, the union between the upper and lower parts of the bowel is not complete, and escape of fæcal matter takes place into the peritoneal cavity: and not unfrequently after the detached portion of bowel has been discharged *per anum*, and the patient promises to make a fair recovery, the seat of separation becomes more and more narrowed, and ends by becoming a tight stricture.

Of thirty-five cases of discharge of bowel *per anum*, collected by Dr. Thomson,[1] sixteen appear to have recovered perfectly, and nineteen died after a longer or shorter interval; and out of nineteen cases, collected by Dr. Peacock,[2] in which the result is mentioned, nine made a good recovery, five still suffered from symptoms indicative of obstruction, and five died subsequent to the discharge of bowel, at intervals varying from forty days to thirteen years. With regard to the period at which the separation takes place, it appears, from Dr. Peacock's paper, that in several cases bowel was discharged on the sixth or seventh day after the beginning of the disease; that in most the discharge took place before the twentieth or thirtieth day; and that occasionally the bowel was not passed until after a few months or even one year had elapsed. In one case fragments of bowel were expelled at intervals during a period of three years. Lastly, in reference to the portion of intestine which thus escapes, it appears that out of forty-three of the cases cited by Drs. Thomson and Peacock, in thirty-two it consisted of small intestine alone, and in eleven only comprised a part of the larger bowel.

(*b*) *The Symptoms* which attend intussusception are made up partly of the symptoms of intestinal obstruction, partly of those of enteritis; but they present much variety, and are often so vague as to render, for a time at least, accurate diagnosis impossible. There are nevertheless certain characteristic symptoms, which if present point pretty certainly to the existence of the lesion in question.

The commencement of intussusception is attended with sudden and more or less severe abdominal pain of a griping or twisting character, which is referred usually to the neighborhood of the umbilicus. This generally ceases after a short time, perhaps a few hours, and then after an interval of comparative or total ease returns temporarily, and thus perhaps continues to recur remittently. There is not necessarily any abdominal tenderness, and indeed the patient frequently finds relief, as in colic, by various contortions of the body and by pressure upon the abdominal parietes. Sympathetic vomiting may be an early symptom, but is often in the beginning absent. Constipation generally follows upon the sudden attack of pain: not however immediately, for the bowel below the seat of lesion may, and does generally, continue to act upon its contents until they are completely expelled: nor necessarily, because, as has been pointed out, the intussusception does not in all cases entirely prevent the passage of fæcal matters from above; and sometimes, indeed, instead of any tendency to constipation there is actual diarrhœa. There is one peculiarity, however, in connection with the intestinal evacuations, which is rarely absent; it is, that very soon after the occurrence of intussusception, the blood which escapes from the deeply congested mucous surface of the invaginated bowel mingles with the contents of the bowel below, and escapes with them by stool in greater or less abundance.

[1] Dr. Peacock's paper : Path. Trans. vol. xv. p. 113.   [2] Ibid.

The symptoms which mark the subsequent progress of the case depend partly on the situation of the intussusception, partly on the degree in which the bowel is strangulated. It has been shown that when the intussusception involves the large intestine, actual strangulation occurs somewhat rarely, and the case tends to become much protracted. In this event the symptoms are apt to be very ill-defined: the paroxysms of pain are often slight, and recur at distant intervals; constipation may exist at the beginning only, or may occur from time to time, or it may never be distinctly present; there is generally more or less vomiting. As the case, however, progresses, the pain often increases in severity; the vomiting becomes more and more incessant and possibly stercoraceous; the alvine evacuations either continue to pass or become re-established, blood and mucous are discharged in variable quantities, and even dysenteric diarrhœa comes on. And then after a longer or shorter period, sometimes after two, three, or four months, the patient, who has been gradually getting more emaciated and feeble, dies of simple exhaustion. When the invagination occupies the small intestine, strangulation is usually of rapid occurrence, and its occurrence adds to the symptoms of mere intussusception those of enteritis. The case, therefore, speedily assumes a very threatening aspect. Febrile symptoms manifest themselves, the abdomen becomes tender, incessant vomiting comes on, and the bowel becomes obstructed, or at all events discharges only those matters which the congested and gangrenous tissues pour out. Under such symptoms, the patient, as in uncomplicated enteritis or internal strangulation, may speedily succumb; but sometimes, at a moment when the disease appears to be still progressing unfavorably, the constipated bowel begins to act, offensive stools mixed with blood and mucus begin to be discharged with more or less tenesmus, vomiting diminishes or ceases, febrile symptoms abate, and after a longer or shorter period of dysenteric symptoms a sequestrum is passed *per anum* in the form of a dark fœtid gangrenous mass.

The most characteristic features, amongst those which have been enumerated in the symptomatology of intussusception, are, first, the sudden onset of the malady, with pain and more or less constipation and vomiting; and secondly, the discharge of blood *per anum* which is generally present even from the beginning. But there is a third sign, to which no allusion has yet been made, which is perhaps of even greater importance, namely, the presence of a tumor. It can scarcely happen that any length of a threefold tube of intestine, especially when its layers, one or all, are congested and swollen, can be present without forming a tumor capable of detection by careful palpation through the abdominal walls, provided at least these be not too fat or too rigid, or the bowels generally be not too much distended with gas, or the abdominal tenderness be not too great, to admit of satisfactory examination. The presence of a tumor indeed, especially in the case of ileo-cæcal, or colic, invagination, may often be recognized during life; and that the tumor is an intussusception may also often be recognized, partly by its cylindrical form, partly by its position, but especially by the fact that it may in some cases be detected changing somewhat from day to day in form and direction, as the intussusception increases, and may sometimes also be felt to dilate and harden, and then subside, under the influence of its peristaltic movements. Further, in those cases in which the intussusception extends low into the rectum, its lower extremity may be detected with all its characteristic features by the finger inserted into the anus.

It must not be supposed, from the foregoing observations, that there is always a wide distinction between the symptoms of invagination of the small intestine and those of invagination of the large intestine. There is no doubt that the majority of jejunal and iliac invaginations are marked by the violent symptoms and rapid progress which have been assigned to them, and that the majority of invaginations involving the large intestine present less urgent symptoms and assume a chronic character. But undoubtedly in some cases invaginations of the small intestine approximate in symptoms and in progress to those of the large intestine, and in a still larger proportion of cases cæcal and colic intussusceptions are attended from an early period with symptoms of great urgency and prove rapidly fatal. These differences depend apparently on the presence or absence of strangulation, which, as has been shown, may occur in connection with any form of invagination, but which generally occurs early when the small intestine alone is affected, late and perhaps not at all when the large intestine is the seat of disease. And it is important to bear in mind that it is this very strangulation, leading to engorgement, inflammation, and gangrene of the invaginated tract of bowel, which, while it gives rise to the most urgent and distressing symptoms, and not unfrequently induces speedy death, is effecting the separation of the obstructing mass, and thus leading to the only possible solution of the case compatible with restoration to health.

There are several additional points in which as a rule differences available for diagnosis are manifested between invaginations of the small and large intestines respectively. Dr. Brinton has especially dwelt upon them. First, tenesmus is common in invagination of the large intestine, but is not necessarily present, and is generally absent when the small intestine is affected ; secondly, hæmorrhage from the bowel (connected doubtless with the relative degrees of congestion of the invaginated portion of bowel) is much more copious in invagination of the small intestine than in that of the large, and blood may also in the former case be vomited; thirdly, obstruction of the bowels is a more prominent symptom when the small intestine is affected than when the large intestine is affected. The remaining points on which Dr. Brinton insists, namely, the situation of the tumor within the abdomen, and the discovery of the end of the intussusception in the rectum, have been already discussed.

Hitherto it has been supposed that the case of intussusception has been uncomplicated with any other malady; but it must not be forgotten that general peritonitis may come on at any time in its progress, and that it is sometimes induced by perforation of the bowel. The latter event is especially apt to occur at the time of separation of the slough, and necessarily renders a case, already sufficiently precarious, hopeless.

The percentage of deaths in intussusception must be very large; it is very difficult, however, if not impossible, to estimate what that percentage is. The stage at which patients die, and the immediate cause of death present very great varieties. Dr. Brinton estimates that the average duration of cases directly fatal is five and a half days. This estimate may probably be accepted with regard to those cases in which strangulation marks the onset of the intussusception, and generally therefore with regard to invagination of the small intestine; but, as Dr. Fagge points out, it can only be true, in a qualified sense, of invagination of the large intestine,—namely, if we reckon the duration of the case from the first manifestation of symptoms of strangulation, and not from the moment at which invagination commenced, which may have been many weeks previously. In cases in which there is not immediate strangulation, the patient may survive for weeks or months, ultimately dying of exhaustion, or killed by the supervention of strangulation. Even after the slough has been discharged, and the continuity of the segment of bowel above and that below the neck of the invagination has been established, permanent recovery would seem to be less frequent than ultimate death,—death being induced at various intervals afterwards, either by exhaustion or by the effects of stricture of the bowel. Recovery after the separation of a portion of the small intestine seems to be more frequent, both relatively and actually, than after the separation of a portion of the large intestine.

(c) *The Treatment* of intussusception, like the treatment of other forms of intestinal obstruction, must be on the whole negative; or, to be more explicit, the less actively the patient is treated, the more likely is he to have his life prolonged, and ultimately to recover. Here, as in most other kinds of obstructive disease, all forms of purgatives must be eschewed, everything in fact must be avoided which can have the effect of promoting peristalsis; for violent movements of the bowel, independently of any other mischief they may effect, naturally tend to increase the size of the intussusception. Neither must it be forgotten that the special ground on which alone the administration of purgatives may be urged exists less in intussusception than in other forms of obstructive disease; for constipation is rarely complete at any rate for more than a few days. On the other hand, opium is of extreme value for the sake both of relieving the pain due to enteritis, or to violent peristalsis, or both, and of restraining the exaggerated movements of the bowel. Dr. Brinton suggests that belladonna, on account of its relaxing influence on the unstripped muscular fibres, may be given with advantage, either alone or combined with opium. Enemata are often beneficial, partly by relieving the lower bowel, partly, perhaps, by acting as a kind of internal fomentation. They may, however, possibly have another value, at all events when administered in large quantities, gradually and without violence. Thus there is some reason to believe that where the large intestine is affected, the distention caused in the external tube of the intussusception, and the pressure exerted on the invaginated portion of bowel itself by such injections cautiously administered, may in some cases, especially those of recent origin, and where the length of bowel involved is as yet small, avail to effect its restoration. Inflation of the bowel *per anum* was long ago recommended for the same purpose; and of late years this procedure has been revived, and several cases have been recorded in which it seems to have been successful. It is obvious, however, that, as is the case with ordinary enemata, inflation can only be of service when the invagination involves the large intestine, and when it is in an early stage. But in intussusception, as in other forms of disease attended with obstruction, the question of surgical interference is not unlikely to arise—Can any surgical operation be performed with a prospect of benefit? It may be supposed that it would be no difficult matter, after opening the abdominal cavity, to withdraw from its sheath an intussuscepted portion of bowel; and no doubt, if adhesions had not yet been formed, or if gangrene had not yet taken place, the evolution of the intussusception might be effected; yet even then considerable force would have to be applied, especially if the intussusception were large, and much risk of damage would attend the process of retraction. Assuming then that an operation might under certain conditions be attended with advantageous results, the question as to what these conditions are naturally presents itself. Now, considering how acute are the symptoms which attend invagination of the small intestine; how speedily adhesions, gangrene, and separation of the slough begin to take place in it; how difficult it is to feel sure of the nature of the case at that early period when alone an operation would have a chance of success; and moreover how often (comparatively) the patient is restored to health by the spontaneous discharge of the invaginated length of bowel,—it seems scarcely possible to avoid the conclusion that in these cases at least surgical interference should be discarded. But when, on the other hand, we bear in mind that in intussusception of the large intestine, ultimate recovery is exceptionally rare, even after the separation of the invaginated portion of bowel, that this separation is of very infrequent occurrence, and that the invaginated bowel is apt to remain in a fairly healthy condition for weeks, sometimes, after the commencement of the disease, it is obvious that we have here an opportunity for operation and a chance of benefit from it very much more favorable than those which iliac and jejunal intussusceptions offer. And it becomes difficult not to accept the conclusion to which Dr. H. Fagge comes, which is to the effect that it is in these cases, and in these alone, that the question of operating should be seriously entertained.

CONCLUDING REMARKS.—Before finally dismissing the subject of intestinal obstructions, it may be convenient to consider, however briefly, some of the more important points upon which our discrimination of such cases of obstruction as may come before us must mainly depend, as well as some of those points of treatment which have a general value in reference to them.

(a) *Pain* is a more or less general and prominent symptom in all cases of obstruction, but it varies a good deal in different persons, both in duration, character, and severity. It is partly the pain of peritonitis, partly that of colic, and these may be present separately, or variously combined. Hence it can be readily understood, that although in well-marked cases the character of the pain may afford us valuable assistance in determining whether the peritoneal surface is alone diseased, or whether the inflammation affects the inner tunics only of the bowels, or whether it involves pretty equally the peritoneal, muscular, and serous coats; in others it affords no evidence whatever of a trustworthy kind. I have a distinct recollection of one of the most extensive and severe cases of enteritis I ever saw, associated with peritoneal inflammation, which a quite well-experienced medical man regarded almost to the last as one of simple colic. It may be added, that even where there is distinct inflammation of a length of bowel, the pain and tenderness, instead of occurring immediately superficial to the affected gut, are frequently most marked in the umbilical region. This latter peculiarity is manifested not unfrequently in cases of inguinal or femoral hernia, and is, indeed, a not uncommon characteristic of affections of the small intestines.

Painful peristaltic movements coming on in paroxysms constitute one of the most distressing, and at the same time one of the most characteristic, symptoms attendant on obstruction; yet, although they may be present in a marked degree in all forms of obstruction, I agree very much with Dr. Fagge in the belief that they are for the most part most severe and most constant in the cases of longest duration; in the cases, therefore, in which enteritis is either not present at all or occurs late.

(b) *Vomiting* is rarely if ever absent from the various affections now under consideration. In the beginning it is sympathetic only, and in that respect resembles the vomiting which attends many other affections not necessarily involving the gastro-intestinal tract. After a while, however, in most if not all cases, it owns a more direct cause. The bowels above the seat of obstruction become distended with contents, partly with what has Leen taken by the mouth and has been transmitted onwards; partly, as Dr. Brinton justly insists, with the secretions of the intestinal walls; these, by the combined effects of simple overflow, of the peristaltic movements of the bowels and of the pressure exerted on the bowels from without, gain an entrance into the stomach, and then become vomited, constituting what is called stercoraceous vomit. The stercoraceous matter, though never in cases of simple obstruction derived from the large intestine, and probably never directly from the lower part of the small intestine, still acquires a thin pea-soup-like aspect and a fæcal odor, which the normal contents of the stomach never do assume, and which are doubtless the result simply of the long residence of the intestinal contents within the bowels, and of their admixture there with bile and other secretions. Vomiting is generally an early symptom in all cases of intestinal obstruction, and in those of acute progress may continue to the end without intermission. Yet even in some of these it intermits, and may be absent for a comparatively long period. In the more chronic affections its occurrence is extremely variable; but even here vomiting generally becomes more or less constant, and then stercoraceous towards the close of life. There is no doubt that vomiting is an earlier, a less interrupted, and a more severe symptom, in proportion to the nearness of the seat of obstruction to the stomach, and that for this reason it is a more marked accompaniment of obstruction of the small intestine than of obstruction of the large.

(c) *Constipation* is naturally one of the most characteristic phenomena of obstructive disease, and its occurrence is of high diagnostic value; yet it need scarcely be repeated that fæcal matters will often pass with comparatively little difficulty through even a tight stricture, especially one in the small intestine; nor must it be forgotten, that generally at the time at which complete obstruction is established, the bowel below contains a larger or smaller quantity of fæces, which may be removed naturally or by injection, and the removal of which might lead to the belief that no obstruction exists. Scybala may sometimes be seen in the large intestine, after death from complete obstruction of the ileum of many weeks' standing. Nevertheless, constipation of an insuperable character is for the most part an exceedingly pronounced symptom; coming on suddenly, and persisting in cases of internal strangulation, and of the lodgment of gall-stones; coming on somewhat gradually, or at all events with premonitory stages, in most cases of stricture and of compression. In intussusception there is also generally sudden constipation, of various duration, but the invaginated mass (especially when the large intestine is involved) is rarely quite impervious, so that before long a slight transmission of fæcal matter begins, at all events in all chronic cases, to take place; moreover, in cases of intussusception, blood is usually passed by stool at an early period, and more or less continuously throughout their whole duration. The discharge, indeed, from the large intestine assumes something of a dysenteric character, and becomes associated with symptoms in some respects resembling those of dysentery. In intussusception of the small intestine, the discharge of blood is sometimes very copious.

(d) *Tumor and Shape of Belly.*—The belly in cases of obstruction soon becomes more or less tense and tympanitic (unless, indeed, the obstruction be in the upper part of the small intestine) in consequence of the distention of the bowel above the seat of stricture by accumulated fæcal matters and by gas; and in some instances the shape which the abdomen then assumes may aid in the diagnosis of the site of obstruction. Thus, if the rectum were blocked up, distention, though soon extending throughout the whole of the large intestine, would first take place and be most extreme in the sigmoid flexure and descending colon, in the situation of which parts, therefore, some special fulness might be looked for; if the obstruction existed in the transverse colon, some fulness would not improbably be discovered in the right flank, and, according to circumstances, in the position of a larger or smaller portion of the transverse colon, the left flank presenting a comparative absence of fulness, tension, and even perhaps of weight; while, again, if the impediment occupied the lower part of the ileum, the distention would probably be most marked in the mild region of the abdomen. But, as has been before pointed out, the evidence afforded by the general shape, and resistance, and weight of the abdomen must be received with great caution, for the distended bowels very readily deviate from their usual position, and diffuse themselves, as it were, beneath the abdominal surface, displacing, or at least concealing the bowels, which are collapsed and empty. Sometimes, indeed, in distention of the large intestine, the sigmoid flexure extends over the whole front of the abdomen, and with the aid of the other lengths of colon effectually conceals the whole of the small intestine from observation. The presence of an abdominal tumor, as distinguished from mere distention of bowel, is an important element in diagnosis. It need scarcely be said that, in internal strangulation, and in most cases of compression, no tumor is likely to be felt; and indeed in stricture also, unless the stricture depend on some form of cancerous growth, or be associated with the presence of peritoneal cancer, or be in the rectum within reach of the finger, no tumor will probably be distinguished. In cases of lodgment of gall-stones, the lump produced by the presence of the gall-stone might, one should suppose, be not very difficult of detection; but unquestionably in the great majority of them, of those even under the care of thoroughly competent practitioners, no tumor has been recognized during life. Indeed it may be pretty confidently asserted that they are rarely, if ever, recognized. This fact may be due in some degree to the absence generally of very minute manual examination; but it must not be forgotten that the tumor formed by a gall-stone is really not very large, that the swelling of the bowel above the obstruction tends to cause the point of obstruction to recede from the surface, or to mask it, and that tenderness, abdominal fatness, rigidity of muscles, and other conditions, all aid more or less to interfere with successful manual examination. Of all the different forms of obstruction which have been enumerated, intussusception is the one which is most commonly attended with the presence of manifest tumor; but tumor seems to be far more common in connection with intussusceptions involving the large intestine than in that form of the disease which is limited to the ileum and jejunum. It is needless to repeat the characteristic features which such tumors present.

(*e*) *The Condition of the Urine* has been regarded ever since Dr. Bar-low's [1] interesting observations on the subject were published, as some indication either of the seat of obstruction, or of some other conditions connected with the obstruction. Dr. Barlow observed that, in a case of his, in which the obstruction was in the duodenum, there was an almost

---

[1] Guy's Hospital Reports, vol. ii. Second Series.

total suppression of urine; and there is no doubt that in many cases of obstruction high up, the same phenomenon is manifested. He argued that the great diminution of this secretion, in his and in similar cases, was caused by the constant vomiting which is always present in obstruction of the upper part of the small intestine, and by the little available absorptive surface which is presented, combining to prevent the entrance of fluid into the vascular system, and the supply of an adequate amount to the kidneys for the maintenance of their secretion. And he argued further, that the abundant discharge of limpid urine which is frequently observed in cases where the seat of obstruction is low down, is to be explained by the presence of entirely opposite conditions. Further observation, however, seems to show that although there may be a tendency on the whole to a diminished secretion of urine when the impediment is high up, and to an increased, or at all events fairly abundant secretion when the impediment is low down, the urine is in many cases abundant or scanty apparently quite independently of the seat of obstruction. Dr. Brinton, indeed, suggests that the diminished secretion of urine which is frequently met with, and the variability of which phenomenon he fully recognizes, is rather due to a kind of vicarious secretion into the bowel above the seat of obstruction, to which also, rather than to ingesta, he no doubt rightly attributes most of the distention of the bowel and much of the vomit. Mr. W. Sedgwick,[1] however, apparently with more reason argues that the diminution or suppression of the urinary secretion is related to the suddenness and intensity of the symptoms, and is immediately due to the reflected influence of the abdominal sympathetic centres. On the whole, even if we adopt Mr. Sedgwick's views, it may probably be accepted as generally true that diminished secretion of urine—often, however, temporary—attends those cases in which the symptoms are of sudden occurrence and acute; and that a fairly abundant secretion of this fluid characterizes cases which are chronic in their course; and that, mainly on these very grounds, suppression or diminution of urine is far more common in cases in which the small intestine is obstructed, than in those in which the impediment occupies the larger bowel.

(*f*) *The Mode of Invasion* is often of great value in reference to diagnosis. Internal strangulation and intussusception always begin suddenly, with more or less acute and severe symptoms. Obstruction by gall-stones might be expected to be preceded by symptoms indicative of the passage of a gall-stone from the bladder into the duodenum, and by further symptoms arising in the course of its journey to the spot at which it becomes finally arrested; and sometimes, but by no means always, the history of such premonitory symptoms can be pretty clearly obtained. Stricture, on the other hand, and in a less marked degree obstruction from compression of the bowel, are in the great majority of cases preceded for a more or less considerable length of time by symptoms which point to what is going on, and which for the most part have a resemblance to those which attend the fatal attack.

(*g*) *The Duration of Life* after the commencement of symptoms which lead to belief in the presence of one of the maladies under consideration varies considerably in different cases. The continuance of life is compatible with the persistence of mere, though complete, colic or rectal obstruction of several weeks' or even months' duration. But death as a rule supervenes much earlier in proportion as the impediment is situated

---

[1] Med.-Chir. Trans., vol. li.

nearer to the stomach. When, however, enteritis is associated with obstruction, then, wherever the obstruction may be, the progress of the case
is always very rapid, and, dating from the commencement of the enteritic
symptoms, rarely occupies more than a week, often only three or four
days. Hence internal strangulations, obstructions by gall-stones, and
intussusceptions in which strangulation occurs (more particularly therefore intussusceptions of the small intestine), are usually fatal within a
few days after the commencement of symptoms; while obstructions from
stricture or compression, and generally also those from intussusception
affecting the larger bowel, for the most part present a comparatively
chronic progress.

(_h_) _Statistics._—There are certain striking facts deducible from the
statistics of obstructive diseases, which it is always well to bear in mind.
First, as regards age and sex. It is a well-ascertained fact that obstruction by gall-stones always occurs late in life, generally over fifty, and
about four times as frequently in women as in men; it appears also that
intussusception may occur at all ages, and is at all ages somewhere about
twice as common in males as in females, but that of intussusceptions involving the large intestine (which form pretty nearly two-thirds of the
total number of fatal intussusceptions), probably fully one-half occur in
children under seven years of age; it appears further that stricture (if we
omit strictures due to congenital malformation) is a disease of adult life
and occurs indifferently in both sexes. Next, in reference to the portion
of intestine involved. Stricture, as a cause of death, belongs almost without exception to the large intestine, and not only so, but at least three-
fourths of the total number of strictures are situated below the middle of
the transverse colon; compression and traction belong essentially to the
small intestine, and may be regarded, as Dr. Fagge observes, in a practical point of view as the strictures of that tract; internal strangulation
occurs more particularly in connection with the small intestine, or with
the cæcum and sigmoid flexure; gall-stones, with hardly an exception,
become arrested somewhere in the jejunum or ileum; and the large intestine is involved in intussusception at least twice as often as the small
intestine alone. Lastly, with respect to the relative frequency of the
several lesions, it may be well to quote Dr. Brinton's figures, based on
500 deaths from obstruction; according to which it appears that out of
100 cases, 43 are cases of intussusception, 17 are cases of stricture, 4·8
are cases of impaction of gall-stones, 27·2 are cases of internal strangulation (including, however, all those cases which have been here described
as compressions), and 8 are cases of torsion, in regard to which the opinion has been previously expressed that they are simply cases of uncomplicated enteritis.

(_i_) _Finally in respect of Treatment_, there are a few established principles which must guide us in all cases of sudden obstruction of the
bowels, and especially in all cases where that sudden obstruction is attended with symptoms of enteritis. First, purgatives however mild can
do no good, may do immense harm, and must be altogether discarded.
Secondly, opiates and other sedatives must be administered largely, or at
least sufficiently largely to produce some visible effect in relieving pain
and giving rest, and should in most cases be administered by subcutaneous injection. Thirdly, but little food and stimulus should be administered by the mouth, for they are almost always immediately rejected, or
if retained fail to be absorbed, and then add only to the bulk of fæcal
matters distending the bowel above the seat of obstruction, in either case

adding to the patient's distress and tending to hasten death. Food given by the mouth should be in small quantities, fluid, and easy of absorption and digestion. There is no reason, however, in many cases, why we should not endeavor to support the patient's strength by nutritious enemata. Fourthly, operations for the relief of intestinal obstructions are rarely followed by satisfactory results; nevertheless, if there seem a chance, however remote, of lengthening the life of a patient who is otherwise doomed to speedy death, few would hesitate to catch at that chance. In some forms of obstruction an operation must from the very nature of things be at least useless, as for example in simple enteritis, in torsion, in most cases of compression of the bowel, and in the impaction of gall-stones; but there can be no doubt that if an operation were performed at an early date, internal strangulations might be relieved with fair success, and intussusceptions might be retracted with frequent benefit. Dr. Fagge is doubtless judicious in recommending an operation for the retraction of ileo-cæcal intussusception, for reasons which have been given previously; and there can be no doubt that if the evidence points at all strongly to internal strangulation, an early resort to surgery should be had. It need scarcely be insisted on that no patient suffering from sudden obstruction with enteritic symptoms, in whom an external hernia, whether strangulated or not, exists or has existed, should be allowed to die without undergoing an exploratory operation at the seat of hernia.

4

ULCERATION OF THE BOWELS.

By John Syer Bristowe, M.D., F.R.C.P.

ULCERATION of the bowels, using the word in its widest sense to indicate all those cases in which the mucous membrane is partially—no matter
how or why—destroyed, is a lesion of very common occurrence, sometimes
induced by the extension of disease from the exterior of the intestine,
more commonly the result of morbid processes commencing in its mucous
and sub-mucous tissues.

I. PATHOLOGY.—(a) *Ulceration beginning from within.*—Ulceration
which originates in connection with the mucous membrane may be found
at any part of the intestinal tract; but there are certain situations in
which it is met with much more frequently than elsewhere: these are the
duodenum, the ileum (especially towards its outlet), the cæcum, ascending
colon, sigmoid flexure and rectum; in other words, the commencement
and the termination of both the larger and the smaller bowel.

The causes of ulceration are very various, and are not always easy to
define, and still less easy in practice to recognize. Some forms of it are
no doubt distinctly the result of the liquefaction or destruction of some
specific deposit, as in enteric fever and in tuberculosis, and perhaps, in the
latter stages of syphilis; and some, as possibly the dysenteric, are due to
some specific kind of inflammation. But in a considerable number of
cases the causes of ulceration are local; the bowel is wounded by some
sharp body which has been swallowed, or is rubbed and irritated by some
partially arrested solid mass, or is fretted by the constant passage over it
of acrid fluids, or presents some localized point or points of inflammation,
which own no more manifest cause than does a pustule of impetigo, a
bleb of pemphigus, or an ordinary boil. It may, however, be conceded,
that even in these latter cases the general condition of the patient has
often much to do, at all events indirectly, with the production of the
ulceration: that, for example, on the one hand the fluids which irritate
are often irritating in consequence of being unhealthy; and, on the other
hand, the fretted bowel often inflames or ulcerates under their influence,
because it was previously congested, or its circulation was sluggish.

Many forms of inflammation of the skin are attended with an excessive
production of epidermis, or with the exudation of matter into or beneath
the epidermis, and thus become characterized by the development of
squamæ or of crusts, on the removal of which a more or less raw surface
is left, and beneath which ulceration is apt to take place. The varieties
of cutaneous inflammation, here very briefly indicated, are for the most
part easy of separate recognition, yet they not infrequently merge one
into the other. But on mucous surfaces the distinctions between scaly,
vesicular, and even pustular affections are rarely, if ever, very obvious, the

delicacy and moisture of the epithelium interfering alike with the formation of a mere dry scale and with the limited accumulation of fluid beneath it. I have used the term " croupous " on another page, to indicate
those forms of intestinal inflammation in which the mucous membrane is
found covered with an opaque adherent film, composed of corpuscular
elements, derived partly from its surface, partly from its glandular invo-
lutions; but I have used it in no specific sense, and believe that, in many
cases at least, the film, or false membrane, is homologous with the scurf
of pityriasis, the scales of lepra, or the vesicles of eczema. Ulceration of
the bowels not infrequently commences with " croupous " inflammation: a
linear or irregularly polygonal or stellate patch of more or less intense
congestion and tumefaction makes its appearance, which soon becomes
covered (excepting, perhaps, at the edges by which it may be extending)
with an opaque whitish or buff-colored exudation, which is somewhat
friable and granular on the surface, and extends by rootlets into the Lie-
berkühnian follicles; the patch of exudation after a time separates, and
leaves sometimes a sound surface, sometimes a slight excoriation, or even
a distinct ulcer, manifested by a somewhat cupped grayish or yellowish
surface and a well-marked margin of congested mucous membrane. Ulcers commencing thus may be met with in any part of the bowels, but
are much more common in the large intestine than elsewhere. In the
small intestine they chiefly affect the free edges of the valvulæ conniventes, and in the large intestine either the projecting ridges formed by
the intervals between the sacculi, or those which correspond to the longitudinal muscular bands. They are very apt to occur, particularly in the
large intestine, in the course of pneumonia, and in cases in which the
patient is dying from many forms of chronic disease, such as Bright's disease of the kidneys, cirrhosis, cancer, chronic phthisis; and, from the
peculiar position which they occupy, there is reason to believe that they
depend, partly at least, on irritation by the intestinal contents. Occasionally we find large tracts of bowel more or less deeply congested, and
studded with irregular patches or bands, or an imperfect network, consisting partly of croupous exudation, partly of consecutive ulceration.

In other cases ulceration commences either from distinct mechanical
injury or from more gradual erosion; the ulcer then being roundish, or
more or less irregular in form, varies in size, presenting a more or less
congested and well-defined, but not necessarily thickened, margin, and a
more or less irregularly excavated shreddy grayish surface. Such ulcers
may be observed when gall-stones or other solid bodies have lain for some
time in contact with a portion of intestinal surface; they occur also in the
large intestine, when it has been long distended with accumulated fæcal
contents. In several cases of long-continued constipation, I have seen
the mucous surface of the larger bowel studded with tracts varying from
about one to twelve square inches in area, consisting of groups of circular
ulcers of the kind now under consideration from half an inch downwards
in diameter, and separated from one another by a network formed of congested and partly undermined bands of mucous membrane.

Sometimes, again, ulcers obviously originate in patches of sub-mucous
suppuration, as we see occasionally in pyæmia, or in patches of sub-mucous slough, like an ordinary furuncle. Among these may, perhaps, be
reckoned the ulcerative inflammation of the follicles of the colon, which
Rokitansky describes, and which seems by many to be considered the
earliest stage of dysentery. The follicles first enlarge to between the size
of a tare and a pea, and become surrounded by a dark red halo of conges-

tion, and then, undergoing suppuration, discharge their contents into the bowel by an ulcerated opening, which eventually enlarges, and forms a circular ulcer with overlapping edges. When the follicles are widely affected, the mucous membrane presents in the first instance a generally congested tuberculated surface, upon which, after a short time, groups of small tolerably deep circular ulcers make their appearance.

In other cases, again, ulceration is produced by the separation of a slough. In various parts of the small intestine, but perhaps most commonly in the duodenum and jejunum as well also as in the œsophagus and stomach, circumscribed patches of intense congestion or of extravasation of blood appear in the substance of the mucous membrane, the patches shortly dying, and coming away either bit by bit or in mass. The formation and separation of such patches are often effected with little obvious change in the parts immediately surrounding them; there is often no unwonted congestion observable, and the pits which are formed by their removal for the most part speedily become effaced. I believe they are most commonly seen in cases of small-pox, typhus, and other such diseases. A somewhat similar condition is sometimes observed in the valvulæ conniventes, and still more frequently in the transverse projections from the interior of the larger intestine, the free edges of which then present a line of ulceration, which looks as though it had been formed by a mere splitting of the diseased mucous membrane, and presents either an ashy or a yellow flocculent surface.

But sloughing to a much more serious extent is sometimes met with, especially in the large intestine; patches of surface become livid, or brown, or nearly black with congestion, and then their central region assumes a gray or ashy color, gets shrunken, depressed, and softened, and soon breaks down into a soft shreddy substance, which partly becomes detached and partly adheres to the floor of the excavation, and to the not yet broken-down edges, which latter tend to spread, and to involve more and more of the surrounding tissues. Occasionally extensive tracts of the mucous surface of the large intestine are covered with sloughing patches, originating in the manner just described.

It is not pretended that all non-specific ulcers arise in one or other of the modes here enumerated, or that the several varieties enumerated are even in the beginning in all cases essentially distinct from one another. Still less do they necessarily maintain these distinctions in the later stages of their progress. Fully formed ulcers indeed present a considerable variety of appearance, dependent mainly on the processes which are taking place in them. Thus, when they are in process of healing, we find the general surface smooth and clean, or it may be granulating, the edges little if at all thickened or congested, perhaps puckered, and sloping more or less obviously to the surface of the ulcer with which they are continuous; when they are sluggish, the edges are more or less tumid and rounded, and it may be overhanging, and the general surface smooth, or somewhat irregular and flocculent; and again, when they are spreading, the surrounding mucous membrane presents more or less intense congestion and swelling, and the immediate edge of the ulcer is either flocculent and ash-colored, or presents a vivid red, raw, bleeding wall, or forms a more or less complete rim of distinct gangrene. The floor of an intestinal ulcer is generally constituted by the sub-mucous tissue, but not infrequently the transverse muscular fibres are distinctly exposed, especially in an ulcer which is still spreading; and when the ulcer tends to perforate the bowel the muscular coat itself becomes opaque, eroded, and in parts destroyed.

The account just given applies to individual ulcers. But very frequently, and much more frequently in the large than in the small intestine, numerous ulcers are present at the same time, and tend to increase either in number or size and to coalesce in a greater or less degree; and then, according to the stage to which the ulceration has advanced, we meet in different cases with either a number of roundish ulcers separated by an imperfect network of mucous membrane, or interlacing networks of ulceration and of mucous membrane, or islets of mucous membrane in an expanse of ulceration; or lastly, extensive tracts from which the mucous coat has been wholly removed. In these cases the transverse muscular fibres are often freely exposed, and the remains of mucous membrane are red and swollen and rounded, and form tubercular excrescences. The bowel, moreover, is frequently much contracted.

Some of the specific forms of intestinal ulceration have been elsewhere considered. There is only one, indeed, tubercular ulceration which needs anything like minute description here. Still it may be convenient briefly to advert to some of the more important features which do, or are supposed to, distinguish them severally. I am not aware that syphilitic ulceration has been surely recognized in the alimentary canal, except in the neighborhood of its inlet and outlet; intestinal ulceration, however, is often met with in persons who have died when under the influence of the syphilitic virus, and it seems at least reasonable to suppose that in some of these cases the ulceration, even though it presents no visible distinctive mark, owns a syphilitic origin. Dysenteric ulceration occupies the large intestine, and occasionally invades also the lower part of the ileum. The mode of origin of the tropical form of the disease is variously described by many, including the late Dr. Baly, it is considered to arise in inflammation and suppuration of the solitary glands; by others it is believed to originate in a croupous form of inflammation; and no doubt it sometimes commences with intense general inflammation, passing at once into gangrene. But, however it may begin, it tends to the rapid destruction of extensive tracts of mucous membrane, and to that chronic condition of more or less extensive rawness which has been above referred to. In typhoid fever a deposit takes place in the solitary glands, and in Peyer's patches (more frequently in the latter than in the former), which become congested, softened, and form flat wheal-like elevations. At the end of a few days, it may be a week, the bulk of the enlarged gland begins to slough, a line of ulceration forms around the slough, and this latter acquires a peculiar yellow or brownish hue. In a short time the slough separates, leaving a circular or sinuous ulcer with congested tumid edges, and an excavated surface, limited either by the sub-mucous tissue or by the transverse muscular fibres. Then usually the edges begin to resume the normal thickness and color of mucous membrane, and to blend gradually with the contiguous surface of the ulcer, which itself fills up and contracts, and ultimately heals with a scarcely or not at all visible cicatrix. At other times the ulcer remains irritable or sluggish, or spreads both in surface and depth, either by gradual erosion, or by sloughing, or by the phagedænic process. And then sometimes hæmorrhage, sometimes perforation of the bowel takes place. Typhoid ulcers vary in size from about that of a split pea to that of the largest of Peyer's patches. They are always most marked immediately above the ileo-cæcal valve (to which part they are sometimes limited), and extend thence, gradually decreasing in number and size, upwards through the ileum and occasionally the jejunum. They occur in the large intestine in about half the total number of cases, being then of smaller size than those in the ileum, and diminishing in number from the cæcum downwards.

Tubercular disease of the mucous membrane of the bowel is one of the most frequent forms in which the tubercular diathesis reveals itself, and certainly the most frequent cause of intestinal ulceration. It occurs in rather more than one half of the total number of cases of pulmonary consumption, and rarely if ever independently of it; and it is often associated with peritoneal and other varieties of abdominal tubercle. It affects primarily the same structures as are affected in enteric fever, namely, Peyer's patches and the solitary glands; and in the small intestine therefore is always most advanced and most abundant immediately above the ileo-cæcal valve, from whence upwards (although it may extend throughout the ileum and jejunum) it gradually diminishes. It affects the cæcum more than any other part of the large intestine, involving also the ileo-cæcal valve and the vermiform appendage; but it may form patches throughout the whole of the colon. The large intestine and small intestine are affected by it with equal frequency, and they are both affected in combination about twice as frequently as they are each affected separately. The tubercular material is deposited, either in the form of gray granules or of yellow cheesy masses, in the substance of the congested and swollen glands, and generally soon undergoes softening, producing a small pretty deep ulcer with thickened elevated overhanging edges. When several of these deposits have softened side by side, as happens in Peyer's patches, the ulcerated area presents in the first instance a kind of honeycombed appearance, the small ulcers being separated by more or less complete bridles of yet undestroyed and thickened mucous membrane, and the general margin, which is also thickened, presents a sinuous or scolloped outline. Tubercular ulcers generally tend to spread by the successive deposition and softening of tubercles at their edges, the tubercles not being then necessarily limited to the glands; and by this process they often extend over a considerable area. In the large intestine the whole mucous membrane of the cæcum is sometimes thus destroyed, and often very extensive tracts of ulceration are found to stud the surface of the colon at more or less distant intervals. In the small intestine tubercular ulceration has a remarkable tendency to spread in the transverse direction and frequently forms bands from half an inch to an inch or more wide, occupying the whole circumference of the bowel. Many of these are sometimes met with at short distances from one another throughout the greater part of the small intestine. In most cases the ulcers still go on enlarging up to the patient's death, and occasionally they lead to hæmorrhage or to perforation. Sometimes, however, they cicatrize more or less perfectly: some cicatrizing indeed while others are spreading or new ones are forming. But, owing to the extensive destruction which tubercular ulceration occasions, cicatrization is generally attended with considerable contraction; so that sometimes in the small intestine, in the cæcum, or in the colon, the calibre of the bowel becomes in consequence so much diminished as to produce a real stricture. Sometimes, again, tubercular deposits dry up or become absorbed without ever undergoing actual ulceration; and it is not a rare thing to find, in cases of chronic phthisis, both in the large and small intestines, small, irregular elevated patches, sometimes associated with ulceration or the remains of ulceration, which present a dark grayish hue and a cicatrix-like appearance, the surface being studded with small granules, the edges being puckered and prolonged by irregular bands into the membrane around, an appearance having some resemblance to that produced by superficial lupus. The peritoneal surface corresponding to tubercular ulcers of the mucous membrane is generally studded with minute gray granulations and the lymphatics ramifying in the walls of the same part, and those extending between it and the nearest mesenteric glands are often filled with opaque white creamy or cheesy contents. It may be added that extensive chronic ulceration of the large intestine, which has all the characters previously described as belonging to the later stages of dysentery, or of non-specific forms of intestinal ulceration, is often met with in phthisical patients; in whom there is no tubercle in any part of the bowel except the ileum, and where therefore it may be a question whether the ulceration originated directly in the breaking down of tubercle, or whether, as seems most likely, it took its origin in simple excoriation caused by the constant passage of irritating secretions from the tubercular bowel above, just as the mucous membrane of the trachea becomes so often excoriated in the course of pulmonary phthisis.

Many intestinal ulcers doubtless cicatrize and leave behind them no traces of their former existence, or, at most, a smooth depression with puckered edges. In other cases, however, and indeed in a large proportion of them, results of more or less serious importance follow.

Sometimes, where a vast continuous extent of surface has been destroyed, as we see occasionally in the rectum and other parts of the large intestine, the mucous membrane never does become restored; and even in cases where the destruction of tissue has been much more limited, the ulcer may assume the character often presented by the chronic ulcer of the stomach, and be ready, as that is, to break out again and again under apparently the most trivial provocation. But generally when a large ulcer heals wholly or in part, some degree of contraction of the calibre of the bowel is the consequence,—contraction which takes place both in length and in breadth, but which from obvious causes manifests itself most conspicuously in the latter direction. Stricture, in fact, often follows such contraction, but especially, and indeed almost exclusively, when the ulceration which has given rise to it has occupied the whole circumference of the bowel, as it does often in tubercular disease, and always after the separation of a mass of invaginated bowel.

Another very common sequence of ulceration is perforation of the intestinal walls at the seat of ulceration, and the consequent communication of the interior of the bowel either with the peritoneal cavity, or with that of some hollow viscus. The most frequent of these communications is that with the peritoneum. Perforation occurs more frequently in enteric fever than in any other kind of disease, taking place generally somewhere in the lower three feet of the ileum, and rarely in the colon. It occurs occasionally only in the course of tubercular ulceration of the bowel, and then also generally in the lower part of the ileum. It is induced sometimes by the constant fretting kept up by the pressure of some hard irritating body, such as a gall-stone or some other form of intestinal concretion. Sometimes it follows upon the ulceration and softening of the mucous membrane, which attend the undue distention taking place often in the bowel above an impediment. Sometimes, again, it results from the separation of freshly united surfaces, as in intussusception. And indeed it may happen in the course of any form of ulceration, or weakness, whether dependent on mere thinning, or softening, or ulceration, or gangrene. The actual perforation, at least so far as regards the peritoneum, which is always the last part to yield, is due generally, perhaps always, to laceration. And although the result of the lesion is general and, with few exceptions, rapidly fatal peritonitis, the lips of the perforation and the contiguous portion of bowel are almost always found adherent by lymph to some neighboring viscus. Indeed perforation into the peritoneum is sometimes staved off, or wholly prevented, by the previous occurrence of localized adhesive peritonitis. It is by the intervention of such adhesion that a perforating ulcer of the bowel comes usually to communicate with some neighboring hollow viscus. The ulcer, having first eaten its way through the thickness of the parietes of the bowel, next perforates the layer of adhesions, then the walls of the attached viscus; and thus establishes a more or less free passage between them, and permits a more or less ready interchange of contents. Sometimes an abscess-like cavity lies between the two organs which communicate, and forms the medium of their communication. Such communications, though generally perhaps permanent, are not always so; and their closure is effected usually by the retreat of the bowel from the organ to which it is adherent, and the consequent formation of a hollow funnel-like passage between them, which becoming longer and narrower, finally closes at its narrowest end, or that furthest from the bowel. There are probably none of the abdominal viscera between which and the bowels communication may not be established by means of ulceration beginning on the side of the bowel. Thus, not infrequently, contiguous portions of the small intestine are found opening into one another, or small intestine into the transverse or some other part of the colon: and thus the rectum or sigmoid flexure, or even the small intestine, may be found to communicate with an ovary or with the urinary bladder; or the duodenum, and perhaps the transverse colon with the gall-bladder; or the stomach with the transverse colon; or again almost any part of the intestinal canal may open through the abdominal parietes, forming a fæcal fistula, or artificial anus. In some cases the perforating ulceration begins in a diverticulum of the ileum, or in one of the false diverticula occurring sometimes in the large intestine. Mr. Sydney Jones' records a case in which ulceration of a false diverticulum in the sigmoid flexure led to a passage between that part of the bowel and the bladder. The results of some of these communications are perhaps of little importance; other communications, however, are not only of dangerous consequence, but also of much interest. Among these latter are especially communications between the colon and the stomach or duodenum, which lead to the occasional or constant vomiting of actual fæces, and the escape of undigested food into the large intestine; and communications with the urinary bladder, which occasion the escape of flatus and of fæces into that viscus, with other consequences which are easy to foresee.

(*b*) *Ulceration beginning from without.*—Ulceration of the bowel beginning from without occurs generally in connection with some abscess of which the intestine has been made to form a portion of the parietes. The abscess is sometimes distinctly peritoneal; sometimes occupies a viscus which becomes adherent to the bowel at the point where perforation is about to take place. Sometimes the purulent matter infiltrates the cellular tissue of the mesentery or of some other peritoneal duplicature, and thus reaches the intestinal walls. If the external abscess attacks a part of bowel covered with peritoneum, it generally causes the erosion of that membrane in the first instance to a comparatively small extent: then the matter undermines it, and accumulates between it and the muscular coat; soon the muscular coat becomes opaque, softened, and perforated in one

---

or more spots, when again an accumulation of matter takes place between the muscular and the mucous membranes, which latter then forms a larger or smaller hemispherical bulging towards the interior of the bowel, on the convexity of which ulceration soon ensues, and the communication between the abscess and the bowel is completed. Or again, a hollow viscus may open by ulceration into the bowel, having first caused adhesion, exactly in the same way that the bowel opens into other organs. By the processes here indicated, peritoneal abscesses discharge themselves into various parts of the bowel; inflamed ovarian tumors communicate with the rectum, sigmoid flexure, or other parts; an ulcerated gall-bladder, or an abscess of the liver, perforates the duodenum or transverse colon; an abscess of the kidney or other form of retro-peritoneal abscess opens on the one side into the ascending colon and cæcum, on the other into the descending colon, or, by burrowing beneath the peritoneum, reaches the rectum, and perforates that. In a similar way, too, an abscess of the liver, or even an empyema, may empty itself into the cæcum or some other part of the large intestine, in or just above the pelvis.

In a few instances, tubercular deposits commencing at the peritoneal surface gradually invade the whole thickness of the bowel, forming here and there large knots of tubercular infiltration of the intestinal walls, which gradually softening lead to the ulceration of the mucous surface over them, to the formation of a tubercular abscess, and even to a communication between the interior of the bowel and the cavity of the abdomen.

It may, perhaps, be added here, that malignant disease of the bowel not only causes ulceration of the mucous surface, but not infrequently produces perforation into the abdomen, and is, perhaps, the most frequent cause of complex and unusual communications between neighboring cavities, and these and the external surface.

II. SYMPTOMS.—*The symptoms* which ulceration of the bowels produces are so constantly associated with the symptoms of those morbid states of system on which the ulceration depends, and are so frequently mixed up with symptoms due to the various complications which follow upon ulceration, that we have seldom the opportunity of studying them in their simple form; and, indeed, if we omit all reference to the symptoms of its complications, we leave very little to be said upon the symptomatology of ulceration. It may be stated generally, that ulceration of the bowels is attended in the first instance with more or less marked febrile symptoms, which assume, if the disease become chronic, a distinctly and indeed typical hectic character; that the affected bowel is more or less tender on pressure, a character which is especially observable if the ulceration be extensive, or if it occupy the cæcum and other parts of the large intestine; that there is some impairment of nutrition marked by emaciation and debility, and feebleness of circulation; and that there is, above all, something abnormal in the action of the bowels and in the evacuations. The stools in ulceration of the bowels are generally liquid, contain an abnormal quantity of the fluid secretions of the bowels, and not infrequently more or less blood; they are, moreover, often pea-soup-like in color and consistence, and much more fœtid than in health; further, they are usually passed much more frequently than natural, and the patient suffers from frequent colicky pains and from tenesmus. But all these symptoms are liable to much modification, and one or even all of them may be absent. Thus, sometimes ulceration is present, especially if it occur high up in the small intestine, without occasioning any obvious disturbance of the bowels. I recollect very well the case of a man who died from gradually increasing emaciation and debility, with no symptoms sufficiently characteristic to point to any one organ as the seat of the disease, and in whom after death the only visible lesion was pretty extensive chronic ulceration at the upper part of the ileum. The bowels, indeed, may be constipated from first to last, as we now and then observe in cases of enteric fever, and as happened in a case of extensive ulceration of the·large intestine which I have quoted in another article, and in which death, and probably the ulceration itself, were due to simple constipation. Ulceration of the larger bowel is much more constantly associated with the passage of frequent and thin evacuations than is ulceration of the small intestine : these may be purely diarrhœal when the upper part of the large intestine is alone involved, but assume a more and more decidedly dysenteric character in proportion as the ulceration affects its lower part; in which latter condition the evacuations, though frequent and passed with extreme tenesmus, are scanty, mucous, and often sanguinolent, and occasionally only containing a little true fæcal matter. It is in this dysenteric form of disease, moreover, that the evacuations become most offensive, the fœtor being sometimes, even though no gangrene be present, putrid and almost insufferable. Besides the slight oozing of blood which tinges the evacuations in diarrhœa of a dysenteric character, hæmorrhage to a considerable amount sometimes takes place, hæmorrhage which may be continuous or recurrent, and sufficient in quantity to destroy life. This accident is not very infrequent either in enteric fever or dysentery, and occasionally results from the perforation of a comparatively large vein or artery. There is little to add, even in regard to the diarrhœa which attends tubercular disease of the bowels, excepting that as the intestinal disease is mostly a progressive one, the diarrhœal symptoms, having once declared themselves, tend to become progressively more and more severe, and that it is for the most part in those cases of phthisis which are attended with intestinal complication that the emaciation is most rapid and becomes most extreme. This is not the place to discuss the various symptoms which are caused by stricture, and by perforation of the bowel, and by the communication of the bowel with other organs, nor to enter upon the description of those symptoms which attend typhoid or dysenteric ulceration.

III. TREATMENT.—*The Treatment* of ulceration merges in·the treatment of the various diseases with which it is connected, and admits, indeed, of but little independent remark. But putting all its complications out of the question, our aim in the treatment of ulceration would seem to be, first, to promote the healing of the ulcer, and to prevent, as far as possible, the local mischances which are apt to follow; second, to check the abdominal discomfort and the diarrhœa which so rapidly weaken the patient; and third, to support his strength directly by all means at our disposal. Whether there are any medicines which are capable of being made to act directly on an ulcer seated at a distance from either outlet may be a matter of doubt; still, from our knowledge of what drugs are useful in ulcers of the stomach and of the lower end of the large intestine, we are justified at least in hoping that some benefit, however infinitesimal, may result from the employment of the same medicines in the treatment of the deeper-seated disease. On these grounds, bismuth, nitrate of silver, iron, copper, the mineral acids and other remedies, have been frequently employed, and often with apparent benefit. But rest, which is so useful an adjunct in the treatment of so many diseases, is of inestimable value in the treatment of ulceration of the bowels. The violent and frequent peristaltic movements and writhings which the ulcers themselves give rise to, tend obviously to prevent them from healing, and add greatly to the danger of perforation; purgative medicines should therefore be entirely, or at least as much as possible, avoided, and further, the exalted peristaltic movements which attend the disease should be restrained. For this purpose various astringent medicines may be used,—lime, tannic acid, chalk, and vegetable astringents; but far more useful than these, as a rule, is opium, in one or other of its various preparations. There are probably few simple combinations more generally useful than the aromatic powder of chalk and opium, and the compound kino powder. But it is well to bear in mind that opium cannot always be taken in these cases. Chronic ulceration of the bowels is often attended with an irritable condition of the mucous membrane of the mouth and stomach, manifested by dryness, soreness, and, perhaps, cracking of the tongue, and heat at the stomach, with nausea—conditions which the use of opium unfortunately often intensifies. If opium then cannot be administered, astringent medicines with carminatives must be alone employed; or some other form of sedative, such as hyoscyamus, belladonna, Indian hemp, hydrocyanic acid, &c, must be resorted to. Opium may often be given with advantage in the form of suppository or of enema. It need scarcely be added that it is never desirable by these means to produce prolonged constipation; and that to obviate this contingency, either the medicines which have produced it must be left off or given in diminished doses, or simple enemata must be employed. It is obvious that the various measures which have just been enumerated, while they check peristalsis, act with equal efficacy in fulfilling the second indication of treatment,—namely, the arrest of diarrhœa. Our third and last object, the maintenance of the patient's strength, must be attained by the exhibition of tonic medicines, and the careful administration of food and stimulants. The form of tonic to be given must obviously be made to accord with the treatment selected to restrain peristalsis and diarrhœa; it must also be adapted to the condition of the patient, as regards his general health and his digestive functions. In the same way the diet must be regulated : nothing should be permitted which is known to disagree with the patient; everything should be well cooked, well masticated, and easy of digestion, and food should be given in moderate quantities, and at regular if not frequent intervals. Farinaceous foods are in many cases most suitable, but eggs, fish, and fowl may often be used with great advantage. Butchers' meat is sometimes wholly inadmissible. For stimulants, nothing, perhaps, is better, in a general way, than brandy and water, sherry, or madeira.

For reasons which are sufficiently apparent, and which have indeed been already indicated, the remarks on the treatment of ulceration are intentionally meagre, and point rather to general principles than to details.

CANCEROUS AND OTHER GROWTHS OF THE INTESTINES.

By John Syer Bristowe, M.D., F.R.C.P.

(1) *Cancerous disease*, to any serious extent, much more rarely affects the intestines than the stomach, and thé small intestine much more rarely than the large. Of all parts of the intestinal canal, the rectum seems to be the most frequently thus affected, the sigmoid flexure next. Yet the bowels are very often the seat of a trivial amount of cancerous deposit; for peritoneal cancer, which is a not uncommon form of disease, is almost always attended with more or less involvement of their serous surface. Cancer rarely originates in the substance of the intestinal walls; but involves them by extension from the serous membrane, from the mesenteric and other abdominal lymphatic glands, from the connective tissue of the lesser omentum, venter ilei, or pelvis, or from the stomach, or the pelvic genito-urinary organs, especially the uterus and vagina. When commencing from the peritoneum, it makes its appearance in that membrane in the form of lenticular or tubercular elevations, which tend to increase in number and to enlarge, and then to coalesce, so as to form a tolerably smooth or somewhat nodulated lamina of various thickness. Generally the cancerous deposits appear first, and are most abundant in the vicinity of the lines along which the peritoneum leaves the bowel; and whether the disease begins in the peritoneum or in the substance of the mesentery and similar processes (but especially in the latter case), the sub-serous connective tissue becomes largely infiltrated and thickened, and the bowel firmly fixed to it as it were in it. It is naturally in the loose tissues around the lower part of the rectum, the cæcum, and the duodenum, that the development of sub-peritoneal cancer is most abundant; and sometimes these parts are thus reduced to mere channels, excavated, as it were, in the substance of a solid mass. Cancerous disease of the outer surface of the bowel may be almost universal; or it may affect tracts of bowel of various lengths; or, again, a band of cancerous deposit may encircle the bowel at some point (generally, in this case, the lower part of the large intestine), while merely a few isolated cancerous nodules are scattered at distant intervals over other parts of the peritoneum.

Cancer beginning on the outer surface tends no doubt, sooner or later, to invade the tissues internal to it; but although there is certainly a great tendency in it to spread laterally, it is remarkable how frequently, even in extensive peritoneal cancer, the muscular and mucous coats escape. When the disease extends inwards, growths of cancer, continuous with those placed externally, perforate the muscular coat, which generally becomes at the same time increased in thickness and marked with vertical bands, of which some appear to be simply fibrous. Subsequently the disease invades the sub-mucous tissue, in which it spreads both laterally and vertically, forming a more or less well-defined, rounded, or nodulated tumor, at first beneath the mucous membrane which is still movable over it, then involving that membrane, and rendering it smooth and fixed. At this stage nodules of cancer, having no apparent continuity with preexisting cancerous masses, are apt to appear in the substance of the mucous membrane. Then soon ulceration takes place, which is sometimes preceded by the formation of a kind of false membrane on the diseased surface, and is often attended with more or less sloughing of the cancerous mass. The diseased tract thus becomes excavated, and then presents either a hard, smooth, cupped surface, or one in which fungous granulations are intermixed with sloughing hollows; the edges being thickened, and either callous and tolerably smooth, or sprouting out with cancerous excrescences.

The direct ill-effects of cancer of the bowels are various. In some cases, especially when the mucous membrane is involved in some considerable area, diarrhœa of a more or less uncontrollable character contributes to hasten the patient's death; in other cases, and generally when the large intestine is the seat of disease, and a limited portion of bowel only is involved, stricture takes place; in other cases, serious or fatal hæmorrhage arises, either from the general surface of an ulcer, or in consequence of the erosion of some large vessel in the progress of the ulceration; and in other cases, again, the bowel opens into the peritoneum, and extravasation of its contents and peritonitis ensue, or communications take place between it and other portions of bowel, or other organs, giving rise to special symptoms of more or less urgency and danger.

The different kinds of cancer affect the bowels in much the same proportion as they affect the stomach; and present, as they do in the latter organ, certain specific peculiarities which may be briefly adverted to. Scirrhus tends to produce contraction of the parts which it involves, and is especially that form of cancer which causes stricture. The ulcer which it yields is very often smooth and excavated; but sometimes, when scirrhus extends from the outer part of the bowel to the mucous membrane, it assumes in the latter situation the character of soft cancer, and forms there projecting growths, or an ulcer with a tendency to sprout. Encephaloid cancer presents various degrees of softness and vascularity, and rarely causes obstruction of the bowel, except by the formation of a tumor, or series of tumors, springing from its mucous aspect and projecting into its cavity. The tumors are rounded, or lobulated, or even villous, and have a great tendency to ulcerate or slough, and bleed. The melanotic variety of encephaloid rarely affects the intestines except secondarily, and in the form of minute discrete black spots, scattered for the most part over the peritoneal surface. Epithelial cancer occasionally involves the rectum by spreading to it from the uterus and vagina; and occasionally, also, arises independently in the lower part of that tube. I am not, however, aware that it ever originates, or is indeed found, in other parts of the intestinal canal. Colloid cancer, or (if it be preferred) colloid disease, affects the bowel usually like scirrhus and encephaloid, from the peritoneal surface, and gradually, like them extending through the intestinal walls, spreads pretty widely in the substance of the mucous membrane, at the surface of which it appears in the form of groups of minute vesicles, reminding one of patches of herpes or of eczema, or (if the fibroid element be in excess) in the form of whitish wheals not unlike those of scirrhus. These become eroded, or more or less excavated, but remain pretty smooth, and secrete in abundance the transparent glairy fluid, with which the interstices of colloid material are filled. Colloid cancer comparatively rarely involves the mucous membrane of the bowel, at any rate to a serious extent. It sometimes appears in the cæcum, sigmoid flexure, or rectum, as a primary disease. Mr. W. Adams[1] records a case in which a colloid tumor, as large as the fist, springing from the posterior part of the rectum, projected into it, and caused symptoms of stricture.

It is difficult, if not impossible, to discuss the symptoms and treatment of intestinal cancer apart from the symptoms and treatment of abdominal cancer generally, or from those of cancer of the stomach and rectum, or from those of its chief local consequences,—namely, obstruction and perforation; it is, moreover, needless, for these are all considered at length in other articles.

(2) *Fibroid infiltration and thickening*, identical with the fibroid form of so-called "scirrhous" pylorus, is met with occasionally in the bowels, where also it constitutes one form of "scirrhus." Its chief, perhaps only, seats are the sigmoid flexure and rectum, where it produces results resembling in almost every particular those which have been described as belonging to true scirrhus. It seems, however, to differ from that in its purely local character, in the absence of all secondary deposits, as well as in its elementary constitution.

(3) *Villous growths* are of occasional occurrence in the large intestine, particularly in the sigmoid flexure and rectum. They generally occupy a limited and well-defined area, which sometimes amounts to three or four square inches or more, and sometimes encircles the gut. The portion of the parietes corresponding to the villous surface is always infiltrated and thickened to a greater or less degree with a kind of fibroid material, which forms the basis from which the villous excrescences spring. The mucous coat and sub-mucous tissue are the parts principally thus affected, and sometimes indeed grow out into a tumor with a constricted neck. The villi are abundant and close-set, easily distinguishable, especially if the tumor be floated in water, often of considerable length, conical, cylindrical or club-shaped, and branching. As we have already seen, villous outgrowths are sometimes distinctly cancerous; but certainly most of those which have been met with in the large intestine seem clearly to have been of a benign character. The presence of a villous tumor sometimes causes hæmorrhage from the bowels, or dysenteric diarrhœa; but its ultimate tendency seems always to produce obstruction. In most of the recorded cases death has been the result of stricture. Occasionally, when the growth is situated but a short distance from the anus, it admits of removal by operation.

(4) *Polypi*, or outgrowths of a non-malignant character, are not very infrequently discovered post mortem attached to the intestinal mucous membrane, especially to that of the lower part of the ileum, ascending colon, and rectum, and are sometimes present here in vast numbers. They seem generally to resemble ordinary cutaneous fibro-cellular or molluscous tumors, and consist, like them, of an outgrowth of connective tissue invested in a layer of mucous membrane, which still for the most part presents its normal structure. It seems not improbable that they occasionally originate in connection with the edges of ulcerated patches; but they doubtless more frequently arise independently of any discoverable pre-

---

[1] Path. Soc. Trans. vol. i.

existing cause. In an early stage they form mere rounded bead-like excrescences, looking like enlarged solitary glands; but they soon elongate, and generally at the same time increase in some degree in other dimensions. When thoroughly developed, they form for the most part cylindrical outgrowths from about a quarter of an inch to an inch in length, and from the thickness of a probe up to that of a director, with extremities which are sometimes bulbous and cauliflower-like, and then highly vascular, and tending to bleed. Sometimes they occur in groups of two or three, or two or three spring from the same pedicle. In the lower part of the ileum, similar bodies, but of a flatter and more leaf-like character, appear occasionally to be produced by mere elongation of portions of valvulæ conniventes. The polypi which have just been described are, as far as I know, of little or no consequence; they occur in persons of all ages and of both sexes, and do not seem to cause any symptoms. Solitary polypi, however, sometimes attain a large size, and may then produce great inconvenience, if not more serious mischief. Pedunculated fibro-cellular polypi from any size up to that of a small pear are now and then met with in the ileum, and are supposed to occasionally cause intus-susception; their most common seat, however, is the rectum, in which situation they cause irritation of the bowels, tenesmus, more or less copious bleeding, and other discomforts. These solitary tumors are generally pretty smooth, but are sometimes lobulated or even warty, and mostly abundantly vascular on the surface.

(5) *Other growths* in the intestinal walls are of no practical importance; they are rare, are not productive of symptoms, and do not therefore call for description. Among them may be enumerated circumscribed sub-mucous deposits of fat; small cysts in the same situation; erectile tumors (Rokitansky [1] considers the polypi above described as being erectile) ; and glandular tumors (in two cases [2] I have met with tumors in the small intestine which resembled the pancreas accurately in structure). Lastly, it may be mentioned that calcareous matter is sometimes deposited in small masses, either on the peritoneal or mucous surface, or in the substance of the intestinal walls.

---

[1] Path. Anat. Syd. Soc. Trans. vol. ii.
[2] Dr. Montgomery, Path. Soc. Trans. vol. xii. p. 130.

DISEASES OF THE CÆCUM AND APPENDIX VERMIFORMIS.

By John Syer Bristowe, M.D., F.R.C.P.

THE cæcum and its appendix are liable, in a greater or less degree, to all those affections which have been described as incidental to the intestinal canal generally. But while some occur here comparatively rarely, or are of trivial consequence when they do occur, others (owing partly to the connections and position of the organs, partly to their capacity and shape, and partly to their structural peculiarities) involve them with exceptional frequency, or induce results which are characteristic either in their gravity or in some of the other features which they present.

I. GENERAL ACCOUNT OF DISEASES OF CÆCUM AND APPENDIX.—Inflammation in its simpler forms affects the cæcum at least as frequently as it affects any other part of the gastro-intestinal mucous membrane. Dysenteric inflammation is only less common here than it is in the rectum and sigmoid flexure. Ulceration of a non-specific kind is perhaps more often met with in the cæcum than in any other named tract of bowel. The ulceration of enteric fever is always more extensive and more advanced in the cæcum than in the colon or rectum, and occurs in it about half as frequently as it occurs in the ileum. Tubercular disease, which affects the large and small intestine with equal frequency, is also generally more severe in the cæcum than in other parts of the large intestine. Cancerous diseases are not very uncommon in this part. And again, the degenerative results of chronic inflammation, and of lardaceous and other forms of deposit, and polypoid growths, occur equally in the cæcum and in the colon and lower part of the ileum. The ileo-cæcal valve and veriniform appendix are for the most part involved whenever the cæcum is the subject of any of the morbid processes which have just been enumerated. The margins of the valve are indeed not infrequently destroyed by ulceration. And the appendix especially rarely fails to present more or less ulceration when typhoid or tubercular deposits occur in other parts of the large intestine.

Strictures of the cæcum form (according to Dr. Brinton) 4 per cent. of fatal strictures of the large intestine. Some degree of contraction at this part is, however, a good deal more common than these figures would seem to indicate. The causes of contraction are, cancerous or other deposit or growth in the walls, and the cicatrization which follows ulceration, especially tubercular and dysenteric ulceration. Dilatation of the cæcum occurs casually, as dilatation occurs in other parts of the intestinal tract, from the temporary accumulation of fæcal matters, or flatus, or both. And it occurs also, as in other situations, as a result of obstructive disease in some part of the bowel below it. In this case the dilatation may be-

5

come very great; and according to circumstances the parietes may be thinned or hypertrophied. It is a point of some importance that not infrequently, even when obstruction is pretty low down, the cæcum is more largely dilated than the length of bowel between it and the seat of obstruction.

Perforation of the cæcum is far from uncommon. Sometimes this ensues on long-continued distention, either from thinning, softening and sudden laceration, or from the ulceration which so frequently attends distention. Sometimes it is caused by simple perforating ulcer, or by the irritation of some foreign body which has been swallowed, has traversed the small intestine safely, and has become arrested in the cæcal pouch. Sometimes it occurs in the course of dysentery, enteric fever, and tuberculosis. Sometimes it is a result of cancerous ulceration. And sometimes it depends on diseases outside the bowel, such, for example, as cancer occupying the venter ilei, or the extension of a psoas, renal, hepatic, pleural, or other abscess. Perforation may take place directly into the peritoneum, lighting up fatal peritonitis; or it may establish a communication between the cavity of the bowel and the sub-serous cellular tissue of the venter ilei, or some adjoining part, and lead to the formation of a fæcal abscess; or again, it may cause a communication with some adherent coil of small intestine.

We can scarcely speak of stricture of the appendix vermiformis; yet occasionally, as a result of ulcerative destruction of the mucous membrane or of other morbid processes, the whole organ becomes shrivelled up or atrophied. Dilatation, too, sometimes occurs when its orifice is obliterated or obstructed. Then the appendix becomes elongated and plump (perhaps as thick as the little finger), presents often false diverticula (resembling on a small scale those of a sacculated bladder), and is distended with a glairy transparent fluid, the secretion of the mucous membrane. Again, the appendix is apt to become perforated. This accident may be caused in any of the several ways in which the cæcum itself becomes perforated. It occurs sometimes perhaps as a result of mere ordinary ulceration. Dr. Murchison[1] records a case in which it happened in the course of typhoid fever, but where there was no escape of fæcal matter. Leudet[2] states that out of thirteen cases of perforation of the appendix, which he observed, six were due to tuberculosis. This statement, however, is certainly not in accordance with general observation. The usual cause indeed of perforation is undoubtedly the presence of some concretion which, by fretting the surface with which it is in contact, excites ulceration, to which the perforation is consecutive. Fæces habitually find an entrance into the appendix; but their entrance and escape constitute a normal process on which as a rule no ill consequences supervene. Together with the fæces, however, insoluble bodies of small size—seeds, bristles, pins, pieces of bone, shot—are apt to enter the appendix; and some of these, from their pointed or angular form, or from their size, become retained and cause ulceration. Perforation has been caused by bristles, by pins, and by pieces of bone: and indeed it was formerly generally believed that the foreign bodies causing perforation were all of external origin, and for the most part cherry or date-stones, or stones of a similar character. There seems no doubt, however, that bodies of this

[1] Path. Trans. vol. xvii. p. 127.
[2] Archiv. Gen. Aug. and Sept. 1859, and New Sydenham Society's Year Book for 1860.

bulk rarely find their way into the appendix, and that what have been mistaken for them have been concretions resembling them somewhat in size and shape, but differing from them in origin and in constitution. The concretions generally met with vary from perhaps the size of a small pea to that of a date-stone: they are sometimes of waxy consistence and lustre throughout; sometimes brownish, for the most part fæcal, and laminated; sometimes again composed almost entirely of earthy phosphates; they consist obviously of the admixture, in unequal proportions, of ordinary fæcal matters and of the secretions from the mucous membrane of the appendix, and have obviously formed in the situation in which they are found, either round a nucleus of solid matter which has been first precipitated and concreted there, or round some comparatively small body of extraneous origin. Sometimes two or three of these concretions are present at the same time. Perforation of the appendix occurs at any part, sometimes at or near its base, sometimes at its point or within half an inch of it, sometimes again in some intermediate spot. The resulting orifice varies in shape and size. Perforation may take place directly into the peritoneal cavity, causing generally acute and rapidly fatal peritonitis, sometimes a circumscribed peritoneal abscess; or actual perforation may be preceded by adhesion of the appendix to neighboring parts, and the formation of a limited abscess either among the adhesions or in the surrounding structures.

It may be added here, in order to complete our summary of diseases incidental to the cæcum and appendix: that the most common form of intussusception, and the most frequent in children, is that in which the cæcum is engaged; that the cæcum is occasionally the subject of internal strangulation, and that more frequently its appendage takes part in the production of strangulation of other parts of the intestine; and lastly, that the cæcum and its appendage, together or separately, are not very infrequently contained in an ordinary hernial sac.

II.—ULCERATION AND PERFORATION OF THE CÆCUM AND VERMIFORM APPENDIX.—(a) *Pathology.*—The terms "Typhlitis" and "Perityphlitis," —the former signifying inflammation of the walls of the cæcum, the latter inflammation in the tissues surrounding the cæcum,—are used frequently, though somewhat vaguely and indiscriminately; but I believe are generally applied to those cases in which there is perforative ulceration either of the cæcum or of its appendix, and in which, therefore, there is either limited suppuration in the neighborhood of these parts, or sudden peritonitis. The perforation in the great majority of cases, no doubt, occurs in the appendix vermiformis: sometimes, however, it occurs in the cæcum itself, beginning there generally from ulceration of the mucous membrane, but occasionally from an abscess situated upon its outer surface. The results which ensue have already been briefly enumerated.

In some instances the ulcer perforates that portion of the bowel which corresponds to the mesenteric attachment, or, if occurring elsewhere in the bowel, the area in which perforation is about to take place becomes adherent to some viscus in the vicinity, or to some portion of the parietes of the true or false pelvis. The morbid process may stop at that point; or the escape of fæcal matter and flatus into and among the tissues may lead to the formation of an abscess, with more or less surrounding inflammation and induration. In the latter event the abscess usually enlarges pretty rapidly, and in enlarging takes a course dependent more or less on its original position, in one case descending into the pelvis, and opening perhaps into the rectum, in another passing out with the pyriformis muscle

and presenting in or below the buttock, in another forming a lump in the
groin immediately above Poupart's ligament, or passing along the inguinal
canal towards the scrotum, or along the psoas and iliacus muscles into the
upper part of the thigh.   But indeed, when once an abscess has formed,
although it may tend as a rule to elect one of several courses, there is
scarcely any conceivable direction which under certain circumstances it
may not take.   No doubt it generally presents itself in the groin as a
hardness or lump superficial to the position which the cæcum normally occupies.   An abscess of this kind may empty itself and become healed by
discharging its contents either through the orifice in the cæcum which
gave rise to it, or through an opening at any one of the spots at which,
as has been shown, it may present; or having burrowed largely it may
form a sinus or series of sinuses which never become obliterated.   The
communication between the abscess and the cæcum is sometimes maintained, at other times is more or less speedily obliterated.

In other cases the bowel ruptures directly into the peritoneum, exciting
at once acute peritoneal inflammation.   This may be so severe as almost
directly to prove fatal: but in most cases the patient survives sufficiently
long to allow of the more or less complete obliteration by adhesion of the
general cavity of the peritoneum, and the formation in the vicinity of the
perforated bowel of a circumscribed peritoneal abscess.   It is not improbable that in some cases the perityphlitic abscesses, the course and progress
of which have been already discussed, are really peritoneal abscesses.
And it may be added that the abscesses originally unconnected with the
peritoneum not infrequently open suddenly into it and evoke, as does the
sudden rupture into it of the cæcum or of its appendix, sudden and severe
inflammation there.

The statistics of "Typhlitis," using this term as expressive of all the
morbid conditions which have just been described, are not very easy to
obtain.   But as regards the statistics of that section of typhlitis which
relates to perforation of the cæcal appendage followed by fatal results,
they seem to show very conclusively that this accident occurs chiefly in
early life, and much more frequently in males than in females.   Thus, in
ten cases analyzed by Bamberger,[1] eight were males, two females; eight
were below thirty years of age, two above thirty.   In thirty-two cases
collected by Dr. Crisp,[2] twenty-nine were males, three females; five
were under ten years, thirteen between ten and twenty, seven between
twenty and forty, and seven between forty and sixty.   And in eight
cases recorded in the "Pathological Transactions" since the publication
therein of Dr. Crisp's paper, five were males, three females; and their
ages ranged from thirteen to thirty-four.

The duration of typhlitis must obviously be very various.   When the
perforation takes place directly into the peritoneum, death for the most
part ensues speedily—generally indeed in from three days to a week; life
may, however, even in this case be prolonged in consequence of the formation of a circumscribed peritoneal abscess, to two or three weeks or
more, and it is not impossible that under the latter condition recovery
sometimes takes place.   In seven of Bamberger's cases the duration of
the illness varied between twenty and fifty days.   But when a fæcal
abscess forms in the tissues in the neighborhood of the cæcum no definite

[1] Ueber die Perforation des wurmformigen Anhangs. : Schmidt's Jahrb. 1859, vol.
ci., p. 184.
[2] Path. Trans. vol. x., p. 151.

limits can possibly be assigned to the duration of the case; sometimes the patient recovers pretty speedily; sometimes, the case, having got apparently into a chronic state, proves suddenly fatal with symptoms of peritonitis; sometimes again the patient lingers for months, or even years, with a constantly discharging abscess or a succession of abscesses.

(*b*) *Symptoms.*—The symptoms which attend and indicate typhlitis are mainly either those of acute peritonitis, or those of local suppuration, or a complex of both.   In those cases in which sudden rupture takes place into the peritoneum, there are very often no premonitory symptoms whatever; occasionally, however, some localized uneasiness or pain, due to the ulceration which is taking place, or to some inflammation of the peritoneal surface corresponding to the seat of ulceration, precedes for a longer or shorter time the violent outbreak.   The patient, while in the enjoyment apparently of perfectly good health, and at the moment probably of making some muscular effort, is attacked with sudden acute pain in the region of the cæcum, followed speedily by collapse, and the diffusion of pain and tenderness over the whole extent of the abdomen. The symptoms in fact of acute peritonitis are almost instantaneously set up, symptoms which only differ from those of idiopathic peritonitis in the suddenness of their invasion and the severity of the collapse, and differ in no degree from those which attend rupture of the bowel from other causes, rupture of the stomach, or rupture of the bladder.   It is needless to dwell on the character of the abdominal pain and tenderness, and on the tympanitic condition of abdomen which ensues, on the dorsal decubitus which the patient is generally compelled to assume, on the quickness and shallowness of his respiratory acts, on his feebleness of pulse, shrunken and anxious expression, and for the most part frequent vomitings and hiccough.   But it may be observed, that in spite of, or rather perhaps in consequence of, the unbearableness of his pain, the patient sometimes assumes positions and makes contortions of his body which might seem to be incompatible with the presence of acute peritonitis; that sometimes the peritonitic indications remain pretty strictly limited to the neighborhood in which they commenced, and that very frequently indeed they do not extend above the line formed by the transverse colon; and that sometimes as the case proceeds, even towards its fatal issue, general peritonitic symptoms almost entirely subside, leaving perhaps a distinct fulness and dulness and tenderness, due to the formation of a circumscribed abscess, in or about the right lumbar or iliac, or the hypogastric region.

In those cases in which an abscess forms in the neighborhood of the cæcum, there are in the first instance pain and tenderness in the region of the cæcum, together with rigors and other general symptoms of inflammatory fever.   Generally, too, there is some distinct fulness and tenderness to be felt.   The symptoms indeed are for the most part those which might be caused by suppuration, of whatever origin, occupying the venter of the ileum.   When the abscess extends downwards into the pelvis, or remains deep-seated, the case is naturally obscure.   When, however, it tends to point anteriorly, we find the fulness and hardness become gradually more and more pronounced; the fulness in fact grows into a more or less distinctly hemispherical tumor over which the integuments become œdematous and congested.   Sometimes, even at this stage, the swelling gradually subsides and disappears, owing to the abscess having discharged itself into the bowel; but more frequently it still enlarges and ultimately opens externally, discharging a greater or less amount of fœtid pus, sometimes having a distinct fæcal odor, or even obviously containing fæcal matter and bubbles of gas. It must, however, be remembered, that not infrequently the communication with the bowel has been cut off before the abscess opens externally, and that the absence of ordure or of gas does not necessarily show that the abscess has not commenced in perforation of the bowel. Sometimes the abscess, after having discharged itself externally, gradually fills up, and complete and permanent recovery takes place. Sometimes, after it has healed externally and appears to have been cured, it forms afresh and presents in the same, or some other, situation. In other cases it remains as a permanently open fistula, or as an artificial anus. In these latter cases symptoms of hectic come on, the patient becomes thinner and feebler, and though in some cases life may be prolonged for a considerable period, death generally ensues from gradual exhaustion at the end of a few months, or at the outside a year or two.

There are, however, many cases in which the perforation of the bowel causes abscess in the first instance, and peritonitis subsequently, either in consequence of a fresh intestinal perforation, or of a rupture of the abscess into the peritoneum, or of the mere extension of inflammation by contiguity. These are the cases in which, for the most part, perforation of the cæcal appendix is said to be preceded by premonitory symptoms; and there can be no doubt that it is chiefly by taking these into consideration that cases of perforation of the appendix are estimated by Bamberger and others to have a duration so much longer than we know belongs to mere peritonitis the result of perforation.

It might naturally be supposed that any disease, affecting so important a part of the alimentary canal as the cæcum, would be attended with some disturbance of the functions of that canal. It does not appear, however, that there is any constant disturbance. Sickness is very often entirely absent. Constipation is mentioned as having been present in many cases at or about the time of perforation; but there does not seem to be any definite connection between these two conditions. And diarrhœa not uncommonly supervenes in the course of the disease; but this again would seem to be for the most part a mere accidental phenomenon.

There are many diseases, or incidents of disease, with which typhlitis may be confounded. It may be worth while briefly to call attention to some of the more important of them. Acute peritonitis of idiopathic origin may sometimes, from its suddenness and severity, and from its happening to take the lower part of the abdomen as its starting-point, be thought to have its origin in perforation of the appendix. So also may the peritonitis caused by perforation of the bowel in enteric fever, especially in those cases in which the febrile symptoms are slight and the patient is not compelled to give up work until the sudden rupture takes place. The same also may be said of all those cases in which peritonitis arises from the perforation of a hollow viscus, or of an hydatid or other abscess, from the laceration of the cyst of tubarian or ovarian pregnancy, or from the extension of inflammation from various pelvic organs, especially those of the female. Again, the local suppuration which attends many cases of typhlitis may in some one or other of its stages be easily confounded with abscesses of other kinds, which form in, or find their way into, the region of the cæcum; among which may be enumerated, psoas abscesses, and abscesses extending from the kidney, the spinal canal, and the pleura. It may similarly be confounded with ovarian tumors or inflammation, with cancerous tumors of the venter ilei or glands in the vicinity of the cæcum, and even under some circumstances with aneurismal tumors.

(c) *Treatment.*—The treatment of typhlitis may be dismissed in a few words, not because it is unimportant, but because it resolves itself into the treatment of enteritis and the treatment of a localized suppuration: the former of which has been discussed elsewhere in this volume; the latter of which is mainly a surgical question. As regards those cases in which there is a direct communication between the bowel and the peritoneum, our main reliance must be placed upon opium; which must be administered, partly with the object of relieving pain, partly with the object of restraining intestinal movements and preventing further escape of fæcal matters. For similar reasons, all purgative medicines must be most carefully avoided. In reference to the employment of local measures, such as leeching, fomentation, and the like, no special observations need be made. It is most important of course to administer nourishment and stimulants; and owing to the comparative absence of vomiting, their administration by the mouth can for the most part be much more readily carried out than in cases of enteritis or of obstruction. It is, however, at the same time essential that the bowels should not be overloaded, and therefore that the food which is thus given should be nutritious, capable of easy digestion and absorption, and given in small quantities at frequent intervals. But here indeed, as in many other cases of stomach and bowel disease, it is important to consider how far we may supplement or replace the duties of the stomach and smaller intestine in the absorption of nutriment, by the regular employment of nutritious enemata. When we have to deal with a case of inflammation, circumscribed in the situation of the cæcum, it need scarcely be said that leeching, poulticing, fomentation, and other local remedial measures will naturally be called into requisition; and that, so soon as there are clear indications of the presence of pus, an opening should be made for its evacuation; and that the abscess having been once opened should if possible be kept open, until we have evidence that its deeper parts or ramifications have become healed. In cases of this kind also the use of opium, though not so universally imperative as where there is peritonitis, is generally desirable if not indispensable; and in them also, purgatives, though not perhaps to be absolutely prohibited, should be employed exceptionally only, and with the greatest caution,—indeed there can be little doubt that if constipation be sufficiently obstinate to call for medical relief, relief will be afforded best, and by far most safely, by the use of enemata. Lastly, in these cases, as in all cases where there is abundant and long-continued suppuration and hectic, it is of paramount importance that the patient should be sustained by abundance of nutritious food, that he should have habitually a fair proportion of stimulus, and that the use of tonic medicines, especially vegetable bitters, and tonic treatment generally, should be systematically enforced.

COLIC.

By J. WARBURTON BEGBIE, M.D., F.R.C.P.E.

THE term Colic is derived from the Greek Κῶλον, the colon, or large intestine.

DEFINITION.—The essential character of Colic, as ordinarily understood, is severe pain in the abdomen (in a restricted view, in the colon), augmenting for a time in severity, and then gradually subsiding; occurring in paroxysms, not stationary, but, on the contrary, moving from place to place, accompanied by a sense of constriction and tearing, for the most part also by that of expulsion.

The term Colic is now used in nearly the same way as the ancient writers employed that of Κωλικός. It is, however, abundantly evident that the disease described under that name, by Aretæus, for example, was of a much more serious nature than ordinary colic; it was indeed a frequently fatal disorder. In treating of Colics, Περὶ Κωλικῶν, the learned Cappadocian physician remarks: Κολικοὶ δὴ κτείνονται εἰλοῷ καὶ στρόφῳ ὀξέως. By Linnæus, among the early nosologists, Colic is placed in the class "Dolorosi," and is thus defined: "Intestini dolor umbilicalis cum torminibus." Vogel, using a similar expression to denote the class, explains the disease as follows: "Dolores: Colica, dolor spasticus intestinorum cum obstipatione, nausea, et vomitu." Sauvage more simply and briefly styles Colic "Dolor intestinorum;" and Cullen, correctly assigning the disease a position in the class "Neuroses" of his nosological system, of which "Spasmi" is the third order, has thus described it: "Dolor abdominis, præcipue circa umbilicum torquens; vomitus; alvus adstricta." By French and German writers the terms "Colique" and "Die Kolik" are respectively employed when treating of this disease.

A vast variety of painful spasmodic affections have been described under the name of Colic. Of these it may only be necessary to adduce as illustrations the following: "Colica Hepatica," "Colica Nephritica," "Colica Uterina," as applied to spasmodic pain, sudden in its occurrence, and apparently affecting the liver, kidneys, or uterus. These expressions are eminently faulty, and it is desirable that their use should be entirely abandoned.

It is to the consideration of the true or simple Colic, the "Colica spasmodica" of not a few writers, that the present article will be devoted. "Lead Colic," or "Colica Pictonum," and for which many other synonyms have been employed, will be separately considered, while the occurrence of Colic, or of colicky pains, as a symptom of different abdominal affections, inflammatory and otherwise, will be noticed in the descriptions of these maladies themselves.

SYMPTOMATOLOGY OF COLIC.—As has already been stated in the definition of Colic, pain is its essential and most characteristic feature. This pain is seldom continued or uniform for any length of time, but, on the contrary, is marked by the occurrence of remissions or intermissions, and likewise by exacerbations, which are frequently of very great, even intense severity. So extreme is the pain of Colic at times as to cause persons of heroism to utter loud groans and cries. While the whole abdomen or any part of it may be the seat of suffering, the peculiar twisting pain is specially experienced in the situation of the umbilicus, as Cullen observed: "præcipue circa umbilicum torquens." [1] Great restlessness and frequent turning of the body, changing from place to place, distinguish the sufferer from Colic. He does not rest in bed, but is prone to rise and pace up and down the room; bending forwards, he presses his hands over the belly; and when the pain augments in severity is glad to fling himself on his face on the bed or sofa. Usually, while the pain lasts, the trunk is flexed, the upper part bent forward over the lower. If the patient be in bed and lying on the back, the lower limbs with bent knees are often brought in contact with the abdominal parietes, and are thus retained for some time by his hands. A position of this kind is meant when French writers, in reference to the sufferer from Colic, use the expression, "le malade se pelotonne," the patient rolls himself into a ball. By very firm pressure over the abdomen, as by lying on the belly, the pain is sometimes mitigated or even for a time removed, and this circumstance is of some importance in distinguishing a spasmodic from an inflammatory pain, in so far as the latter is invariably aggravated by pressure.

The form of the abdomen is altered during the continuance of Colic. There may be, and this condition is fully the more frequent, distention, with which there is associated the development of flatus on a large scale, or the parietes of the abdomen may, on the other hand, be retracted. The condition of a distended colon, the seat of pain, may be mistaken for that of gastric distention and pain. When the former, however, occurs, as a phenomenon of the attack of Colic, there are present also other indications of intestinal suffering, such as irregular contractions which may frequently be felt by the hand or seen, borborygmi, and specially the sense of bearing down towards, and constriction at, the anus. Besides, as Dr. Wilson Fox[1] has pointed out, pain arising from the large intestines is seldom felt so much at the ensiform cartilage (the common seat of gastric uneasiness) as in the right or left hypochondriac regions, while there exists a distinct difference between the notes to be elicited on percussion, from the two organs; that from a distended colon being the less prolonged, and having a higher pitch.

Great general depression is capable of being produced by an attack of Colic. This is seen in the frequently pale countenance of the sufferer, whose pulse also is found to be extremely feeble, while the surface of the body is bedewed with a cold and clammy perspiration. The relation of constipation to Colic is most important. A confined condition of the bowels is usually, though not invariably as some writers have asserted, associated with Colic; and not unfrequently, when the bowels have been

---

[1] A recent, perhaps the most recent, French writer on Colic (M. Martineau), in describing the pain, remarks : " La douleur est toute spéciale. Les malades en proie à une colique éprouvent une douleur vive, exacerbante, mobile, ayant une grande tendance à s'irradier. Elle se traduit par une sensation de constriction, de resserrement, de tortillement, ou par une sensation de déchirure et même d'expulsion.—*Nouveau Dictionnaire de Médecine et de Chirurgie pratique*, vol. viii.

[2] The Diagnosis and Treatment of the Varieties of Dyspepsia, p. 53.

efficiently acted on by medicine, the pain, which may have been of the severest type, entirely disappears. Neither is this latter however, the constant result, for, notwithstanding the operation of laxative and cathartic remedies, the pain in some instances proves persistent. Such cases are infinitely less alarming than those in which obstruction of the bowels continues, while the abdominal pain either diminishes or disappears, for in these circumstances the occurrence, sooner or later, of a regular attack of ileus is to be apprehended; while in the former case, the free movement of the bowels, although not immediately, and it may be not even speedily, bringing relief to suffering, is surely succeeded by such before any lengthened period has passed. In some instances of Colic, a confined condition of the bowels is really the cause of the attack of painful spasm, while in others the constipation is the effect of the spasm. In the more protracted cases of Colic, a general febrile state is liable to be induced. Vomiting may accompany Colic, but is by no means a constant or characteristic symptom of this disorder. Much importance is to be attached to the pulse in Colic, for by its condition we are not unfrequently able to distinguish between a simple, although severe spasmodic affection, and an inflammatory disorder. It is to be remembered moreover, that in some circumstances the latter is not unapt to supervene upon the former. Now, in Colic, while the suffering is even intense, the pulse may be little if at all altered. Assuredly it is by no means uncommon to find the pulse under such circumstances remaining tranquil, and in fact altogether normal. Smallness of the pulse, associated with marked depression of the circulation generally, hardness and irregularity, are, on the other hand, of sufficiently frequent occurrence in cases of Colic.[1] The respiration is hurried, and frequently unequal. The voice is apt to be affected in cases of marked severity; it becomes hoarse, while at times it is so enfeebled as to be almost obliterated. The accession of Colic is by no means uniform or exact. The disease may be established suddenly, even abruptly, and without any apparent cause, or it may come on gradually, succeeding the occurrence, for a time longer or shorter, of abdominal uneasiness, and very probably of occasional cramps, which are clearly traceable to some sufficient cause. Not less variable are the progress and duration of the malady. It may exist for days, or last only for hours, or even minutes. These irregularities are largely determined by the precise causes of the attacks. An irregular intermittence is a characteristic feature of Colic; the duration of the painful seizures, and of the intervals which separate them, being subject to great variety.

PATHOLOGY OF COLIC.—Although the relation of the abdominal pain and spasm in Colic to nerve irritation, is obscure, the following remarks appear to be called for. It has been clearly shown by carefully conducted experiments, and is now admitted, that the pneumogastric nerves possess an influence on the movements of the intestinal canal. Such experiments as those referred to have exhibited the contractions of the muscular coats of the intestines under the application of electrical irritation to the vagi, of as rapid and violent a character as those of voluntary muscles, when their motor nerves have been subjected to a similar irritation. Again, when on irritating the ganglionic plexuses surrounding the aorta, by means of the rotary apparatus (*durch den rotatorischen Apparat*), the small intestines and colon, which had been previously wholly inactive,

---

[1] In describing the pulse of Colic, Henoch remarks, " Der Puls ist klein und härt-lich." (Klinik der Unterleibs-Krankheiten.)

when the current began to operate were seized with universally active movements, which continued for a long time after the current was interrupted. It is of further interest to note, that among central portions of the nervous system it is the medulla oblongata which, when irritated by the galvanic current, excites in a decided manner the movements of the stomach and the intestinal canal. Budge saw the same result produced in rabbits, but in a less degree, by irritation of the cerebellum. The spinal cord and cerebrum possess no such influence. All experimenters have described the movements of the intestinal canal as distinctly peristaltic or vermicular.[1] M. Martineau, in his interesting article on Colic to which reference has been made, has pointed out that while the pneumogastric nerve is more especially distributed, as is well known, to the stomach and liver, a portion of the right nerve passes to the semi-lunar ganglia to anastomose with the splanchnic nerves of the great sympathetic, and thus to form the solar plexus. Galvanization of the solar plexus and of the superior mesenteric ganglia equally causes contraction of the small intestine and more rarely of the large. Valentin has made the very important observation that an irritation of the fifth nerve, at the base of the skull, invariably gives rise to peristaltic movements of the small intestine, especially of the duodenum and upper part of the jejunum. Such being proved experimentally, we can understand the occurrence of intestinal spasm or Colic, as the direct consequence of some forms of cerebral irritation. And although, as Romberg has remarked, little is known respecting the influence which is exerted by the affections of the spinal cord and brain, upon spasms of the bowels, the very potent operation of the emotions, fear and fright especially, but in some instances also joy, in increasing the movements of the intestines is thoroughly appreciated.

ETIOLOGY OF COLIC.—Certain temperaments appear to predispose to the occurrence of Colic. Of these the nervous and lymphatic are the most distinguished. Sedentary occupations act in the same manner. The influence of age and sex is sufficiently marked to be worthy of notice. In youth and adult age, Colic is more common than in advanced life, and among females it occurs more frequently than among males. Among the exciting causes of Colic, one of the most frequent is the presence of some indigestible article of food in the bowels. The influence of cold in producing attacks of Colic is also remarkable, and particularly, it has been noticed, cold applied to the feet. There are some individuals who are certain to suffer from an attack of Colic, if by any means their feet have become cold. The association of biliary derangement with the occurrence of intestinal spasm is not uncommon, and this particular form of the disease has been designated "Bilious Colic." Its distinctive features are the vomiting of biliary matters, and the presence of a more or less icteric tint of the conjunctivæ and surface of the body. Lastly, under this head, it is to be held in remembrance that in some instances the existence of a gouty or rheumatic habit of body plays a decided part in the origination of attacks of Colic, although it may probably be admitted that such constitutional disorders are still more potent in determining the true enteralgia or neuralgia of the bowels, a disease which is to be distinguished from Colic.

TREATMENT OF COLIC.—To relieve pain, and generally speaking to act on the confined bowels, are the chief indications for treatment in Colic. In the milder instances of the disease, unaccompanied by any notable de-

---

[1] Romberg, Lehrbuch der Nervenkrankheiten des Menschen ; Darmkrampf.

rangement of the "primæ viæ," this can usually be accomplished by the external application of warmth, or of rubefacients, such as mustard and turpentine, and by the administration of a little stimulant, or carminative mixture. A small quantity of brandy with hot water, a teaspoonful of the compound tincture of cardamoms in warm water, or twenty drops of the compound tincture of chloroform, will be found very serviceable for this purpose. Preparations of peppermint, ginger, and cloves may also be similarly employed. In more severe cases of Colic, or in instances where the remedies already mentioned have failed to relieve the pain, it will be necessary to administer anodyne medicines, and as early as possible to evacuate the bowels. The preparations of opium are most useful among the former; the compound tincture of camphor or English paregoric—in doses of thirty to sixty minims—or a full dose of laudanum. With these a dose of castor-oil, or compound rhubarb powder (Gregory's mixture), should be given, and repeated after a short interval if relief to pain and solution of the bowels be not obtained.

A tablespoonful of castor-oil with twenty-five drops of laudanum in peppermint water, or two teaspoonfuls of Gregory's mixture with a teaspoonful of compound tincture of camphor, and a similar quantity of aromatic spirit of ammonia in a small wineglassful of cinnamon water, will be found most available prescriptions in such cases.

When the attack of Colic has speedily succeeded the taking of some indigestible article of food, it may be advisable to produce vomiting by the administration of an emetic of ipecacuanha wine, or by draughts of hot water.

Should the bowels not respond to the mild remedies already mentioned, it will be necessary to have recourse to the use of stronger cathartics. Of these, sulphate of magnesia, particularly with the addition of a little sulphuric acid, as Henry's salts, and senna, also the compound jalap powder and calomel, may be regarded as the chief.

The employment of laxative enemata should also be had recourse to. A large injection of warm water will frequently be found most useful in relieving the pain, and in effectually acting on the bowels in cases of Colic.

The prophylactic treatment of Colic consists in a careful regulation of diet, particularly in the avoidance of all indigestible articles of food, and in the protection of the surface of the body from the injurious influence of cold. Wearing flannel over the abdomen, and the warm covering of the feet, are especially to be enjoined.

COLITIS.

By J. Warburton Begbie, M.D., F.R.C.P.E.

There seems to be some ground, at all events for supposing that the large intestine may be the seat of inflammatory action, differing in essential particulars from the dysenteric process which will be immediately described. To indicate the simple inflammation of the colon, as distinguished from dysentery, the term *Colitis* has been employed. *Colonitis* has been used in the same sense. The French have the word *Colite*, and the Germans the expression *Entzundung des Schleimhautes des Kolons*.

In dysentery the mucous membrane of the rectum and colon is primarily involved while the pathological changes which are so eminently characteristic of the disease are wrought in it. In Colitis, on the other hand, there is in all probability a commencement of inflammation in the submucous or connective tissue, which underlies the mucous membrane, the glandular structures of the latter being in the first instance uninvolved. The result, however, is a diffuse gangrenous inflammation of the mucous membrane; and when this has occurred, there is no possibility of distinguishing the ulceration thus formed from that which has resulted from the dysenteric process.

It is, however, to be borne in mind that the most experienced physicians and ablest writers have differed in respect to the essential pathology and the characteristic morbid appearances in dysentery. The necessary existence of ulceration has, for example, been denied by some, and the special participation of the glandular structures of the colon, so commonly conceived to hold true of dysentery, has been equally opposed by others. In these circumstances it must be admitted that great difficulty at present exists in the way of correctly distinguishing between the different forms —if there really be different forms—of inflammatory disease affecting the colon, and renewed investigation with careful examination of the various structures and tissues entering into the anatomy of that portion of the intestine, is required before any satisfactory conclusions on the subject can be arrived at.

DIARRHŒA.

By S. O. Habershon, M.D.

Diarrhœa consists in the abnormal frequency of evacuation of the bowels, as defined by Cullen, "Dejectio frequens; morbus non contagiosa; pyrexia nulla primaria:" and it arises generally, but not exclusively, from an irritated condition of the large intestine.

It manifests itself in various forms, some of which have received distinctive appellations, as *Diarrhœa crapulosa, biliosa, mucosa or catarrhalis, dysenterica,* and *choleraica,* to which might be added *nervosa,* and *colliquativa.*

Diarrhœa crapulosa is that state in which there is an unnatural fluidity and excess of fæcal excretion, in which the evacuations are healthy in character, but in excessive frequency and fluidity; in some cases very large quantities are discharged without any discomfort, but, on the contrary, with relief to the patient. This form of diarrhœa should not be checked when it is a natural discharge; but more frequently it is the sequence of irritating and undigested food. Too great a quantity may have been taken, and a portion of it may have passed into the intestine crude and partially dissolved; or from its insoluble character portions of the food, as the woody fibre of vegetables and fruit, may have remained unchanged by the gastric juice, and irritate the intestine. Again, active mental or bodily exercise immediately after a meal, which has been suitable both as to quality and quantity, may interfere with the proper solution of food, and lead to its hasty passage into the duodenum.

When the alimentary canal becomes in this way loaded with undissolved ingesta, pain of a griping and twisting character ensues, from irregular peristaltic action and from distention. The abdomen becomes full; the skin and complexion sallow; the tongue is furred; the pulse is compressible; headache and giddiness are often present; the sleep is disturbed; the bowels act frequently and irregularly, and the motions contain undigested substances, with fluid fæces or with firm scybala. Considerable soreness is at times experienced in the course of the large intestine, and distressing tenesmus arises from the irritation of the mucous membrane of the rectum.

The term lientery is used to designate the condition in which the food is passed almost unacted upon, either by the gastric or intestinal secretions, and in a very short time after having been taken. This state arises from excessive irritability of the whole intestinal tract, with disordered secretions; it is not unfrequent in children after protracted diarrhœa, and gastro-enteritis. It is of common occurrence among the out-patients of large hospitals; and in not a few cases leads to a fatal termination.

Bilious Diarrhœa is also a form of disease produced by the effusion

6

of irritating substances into the intestine; not, however, from without, but from the liver, and possibly from the pancreas and follicular glands. The secretion of the liver becomes either excessive in quantity, or irritating in quality; and the contents of the canal are apparently hurried onward, and evacuated as frequent loose and bilious dejections. The causes of this state are various, and sometimes the disorder of the liver is really secondary to an irritable condition of the intestine itself, due to excess, especially of stimulants. Exposure to cold and wet induces diseases of this kind, especially in the autumnal season of the year. The symptoms are somewhat similar to those previously mentioned; the pain is slight, unless the disease becomes aggravated; the tongue is furred; the complexion is sallow; some febrile excitement is present with frontal headache; pain in the abdomen and in the hypochondriac region. This form of diarrhœa is sometimes epidemic, attacking considerable numbers exposed to similar exciting causes; and when severe, and accompanied with colic or spasmodic pain in the abdomen and legs, and especially with vomiting, it constitutes English cholera, and often leads to great prostration of strength. The countenance becomes haggard, the eyes appear sunken, the pulse is exceedingly compressible and failing, the temperature below normal, the tongue is brown, and the patient too frequently sinks exhausted, especially if very young, or advanced in life, or if already prostrate from other disease.

Abnormal conditions of the bile tend to produce other modifications; thus, the motions in diarrhœa are sometimes in a state of fermentation; they are watery, frothy, and only contain fluid fæcal matter. This I have seen very prominently in a case of phthisis, in which there was probably some ulceration of the intestine, when the evacuations consisted of long shreds of mucus, and casts composed of columnar epithelium and nuclei. After a few weeks this condition subsided under the use of cusparia, sulphuric acid, and opium, with occasional starch injections, but it was followed by very severe pain in the course of the colon, and by frothy, yeast-like evacuations. For this state I used injections of charcoal,[1] ℥ ij. to about a pint of thin barley-water, with great relief; the character of the evacuations improved, and in a short time became naturally fæcal, the pain diminished, and the strength increased. I afterwards gave the patient several grains of myrrh, twice or three times a day, with manifest improvement, till she left the hospital several months later.

Diarrhœa sometimes occurs with an absence of bile in the evacuations; in jaundice this may be the case; it is so in cholera; and towards the close of chronic disease the liver may cease to pour out its ordinary secretion. I have seen it in a patient slowly sinking from the exhaustion consequent on diabetes, without phthisis. The motions were in that case often quite white, like water frothy from an abundance of soap.

There is, also, a form of diarrhœa arising from the inhalation of noxious effluvia, which is closely allied to that just described; the fumes of sulphuretted hydrogen gas are absorbed by the lungs, and through their minute capillaries enter the blood; the gas is circulated and acts as a poisonous agent on that vital fluid, and if concentrated, proves rapidly fatal; if less concentrated, it produces headache, and frequently also diarrhœa. It appears, that not only are the secretions of the liver and alimentary canal changed, but that, by means of this excessive action of the abdominal viscera, the poison is eliminated from the system. So rapid

---

[1] See Dr. Theophilus Thompson's Lectures on Phthisis.

is this agent in its action, that to be present for a short time, even a quarter of an hour, in a dissecting-room, will, in some persons, produce distressing diarrhœa.

In typhoid fever, and in phthisis, ulceration of the small intestine is frequently found to be accompanied with diarrhœa; of these we have spoken elsewhere; in some of these cases the large intestine is involved, but in others, when the diarrhœa has been severe, such has not been the case. It would appear that the continuity of structure with the ulcerated ileum, the irritating excreta, as well as the changed and probably accelerated peristaltic action of the small intestine, tend to excite over-action of the colon, and thus to set up diarrhœa.

Catarrhal and *mucous diarrhœa* arise from a state of slight inflammatory disease, closely allied to ordinary coryza, affecting the mucous membrane of the *large* intestine. The secretion is at first checked, but afterwards greatly increased, and a watery feculent mucus is discharged mixed with the ordinary fæces. This state may continue for several days, or even for a much longer period: the motions are loose, and somewhat watery; and if the rectum be affected, considerable tenesmus is produced; the pain and febrile excitement are slight, but the strength of the patient is reduced, and he is unequal to his usual duties; the tongue is clean, the pulse is compressible; the bladder sometimes sympathizes with this irritation, and a frequent desire to pass urine is induced; in little girls, also, a muco-purulent secretion often takes place from the vulva; redness of the parts is produced with smarting pain, and the idea has sometimes been suggested that the child has been cruelly treated.

In this form of diarrhœa the evacuations contain a considerable quantity of mucus, and a little blood is often observed; these are especially present when irritation occurs very low down in the rectum, or is set up by hæmorrhoids; and the mucus will sometimes pass both before and after the dejection.

In infants the disease closely resembles gastro-enteritis, or it is, perhaps, rather identical with it, but differing in degree, as a greater or less part of the alimentary canal is affected; in these cases the whole tract sometimes becomes rapidly involved, and great, if not fatal prostration, rapidly ensues. (See Muco-Enteritis.)

As with bilious diarrhœa, before mentioned, it is in very young or aged subjects that catarrhal diarrhœa, or catarrhal inflammation of the large intestine, leads to more serious disease, but it is also found among those in whom chronic or more exhausting disease has existed.

This catarrhal diarrhœa not unfrequently becomes a chronic disease, the more severe symptoms cease, but still the bowels do not act in their normal manner; constipation often ensues, and afterwards a fresh looseness of the bowels, and this alternation is oftentimes repeated, or the more solid motions are followed by a discharge of mucus coating the fæces; sometimes the mucus is passed in considerable quantity after the evacuation, or it forms an elongated flake or cast of the intestine. I have observed this condition following severe disease of the intestines of a dysenteric character, and it is sometimes associated with a state of chronic congestion of the liver; again, it is often perpetuated by the presence of hæmorrhoids, and by ovarian disease. It may exist for many years without causing much derangement of health.

Morbid anatomy.—Many instances have been known of fatal diarrhœa in which the appearance of the mucous membrane has been normal, its congestion has entirely disappeared, and a thin mucus only has been found upon the membrane. But this is not always the case, and there are several recognized pathological changes which are frequently present. First of these, because most frequent and therefore the more important, is a vivid injection in more or less isolated patches.

2dly. When the diarrhœa has been chronic, the mucous membrane is not unfrequently covered by a thick layer of mucus, and presents a gray color. I have frequently examined membranes thus changed (as before described; see Duodenum and Cæcum), and have observed that the color arises from minute particles of dark pigmental matter deposited in the substance of the mucous membrane. Prolonged congestion is known to give rise to similar pigmentary changes in many parts, as in the skin, liver, lung, heart, &c., and wherever this pigmentary deposit occurs it is found to be due, as I have described here, to grains of varying tint—orange, red, brown, or black. One must regard these grains as the remnants of actually extravasated blood or to the arrest of some of the oxidizing or other processes which the blood coloring matter probably undergoes in its passage through the various tissues.

In the large intestine this pigmental deposit is found in minute circles around the follicles.

3dly. The mucous membrane, and also the connecting cellular tissue, become thickened.

4thly. Minute ulceration, probably follicular, is found extending through more or less of the length of the colon. These ulcerations are about one-sixteenth of an inch in diameter, and present a minute black zone around each of them. This state would be regarded by many as the result of dysentery.

Dysenteric Diarrhœa.—Purging is the most marked symptom of dysentery, and the lesser degrees of irritation which we have considered under the term of catarrhal diarrhœa might be regarded as a form of dysentery of the mildest character. In dysentery, however, the diseased mucous membrane rapidly passes into a state of ulceration, and blood is discharged with the fæcal excreta.

In *Choleraic Diarrhœa* a thin, very abundant watery mucus is discharged from the alimentary canal. The evacuation may have very little color, and present the appearance of rice-water. It is often alkaline in character, and consists of nuclei and epithelial cells in various degrees of development. After death the membrane is found to be entire, and pale or sodden; the solitary and Peyer's glands are enlarged. In many cases of uncomplicated cholera which I have examined, no further morbid appearance was presented.

Of late years a belief in a fungous growth has been revived, and the dejections of cholera have been said by Hallier and others to contain specific spores. Some very careful and prolonged observations, however, by Drs. Lewis and Cunningham, in India, controvert this opinion.

The symptoms are those of rapid prostration, with pallor and sunken eye; the pulse is compressible, the tongue is cool, and the voice is often scarcely audible; the abdomen is collapsed, and the urine is scanty in quantity; the stomach is often exceedingly irritable, so that everything is at once rejected from it; the alvine evacuations are generally frequent, and of the character before mentioned; and severe cramps in the legs and in the abdomen are often present. This state may pass into one of profound collapse, even after one evacuation of the character of rice-water, but as the prostration subsides, in favorable cases, I have never observed the febrile excitement which is secondary to true cholera.

Another kind of diarrhœa is that which has been correctly called *Serous*, and which is frequently observed in albuminuria. A dropsical condition of the mucous membrane is induced, and the serous exudation from the overcharged capillaries leads to watery discharge into the colon, and thus to diarrhœa. This state of the mucous membrane is precisely analogous to the œdema of the lungs, and to anasarca of the cellular tissue in renal disease. So frequently is diarrhœa present in these cases, that it may almost be regarded as a symptom of the disease, and when moderate is beneficial in its results. It is the action we often seek to produce artificially by powerful hydragogue cathartics, so as to diminish the quantity of urea circulating in the blood, and to relieve the oppressed kidney. All these fluid evacuations contain urea, as does the gastric juice and the mucus discharged from the lungs.

Another class of cases which can scarcely be placed among those previously mentioned, arise from fright, from excessive mental agitation, from want of food, and from exhausting disease; the former cases are of mental origin, the latter constitute what is sometimes called "colliquative diarrhœa;" and the condition of the mucous membrane corresponds to that of the skin, from which profuse partial sweats break out.

In fright the capillaries of the face become blanched, and the blood leaves the whole of the surface; the cavities of the heart are increasingly distended, hence the discomfort there experienced, and the mucous membrane of the intestine is probably also engorged; therefore the discharge from the mucous membrane is to a certain extent beneficial in relieving internal congestion. The intimate connection of the sympathetic nerve with the centres of thought and feeling is the probable explanation of these instances of diarrhœa following mental agitation.

In scurvy, purpura, starvation, &c., the altered character of the blood leads to the effusion of serum, or blood, into the mucous membrane, or into the canal itself, corresponding to the effusion into the skin. In some fatal cases of purpura the whole of the mucous membrane of the alimentary canal is studded with spots of ecchymosis. An interesting case of this kind occurred at Guy's Hospital in 1856, in a young man who had been starved to death.

Discharge of blood, or melœna.—Obstruction of the portal circulation, either from pulmonary, from cardiac, or from hepatic disease, leads to great engorgement of the mucous membrane of the whole alimentary canal; and this congestion may cause hæmorrhage from the bowels. In examining the mucous membrane in these cases, it is very common to find points of ecchymosis, and the capillary vessels of the membrane much distended. Under a low magnifying power we find the capillaries beautifully injected, with extravasated blood between them, still, however, restrained by the unbroken epithelial surface and its basement membrane; if the rupture of this membrane occur blood is extravasated. The discharge of blood may be a symptom of various diseases; thus, ulceration is a frequent cause of hæmorrhage from the bowels, and the ulcer may be located in any part of the canal; in the stomach and duodenum from various causes; in the small intestine in fever and in phthisis; in the colon in dysentery, &c.

The blood does not always present the same appearance; if it arise from hæmorrhoidal vessels the blood will be florid, and precede or follow the dejection; if it come from some higher part of the canal it is incorporated with the fæces; and when it has traversed a considerable portion of the canal, it becomes altered by admixture with the secretions from the mucous membrane. This is the case, to some extent, when the blood is poured into the cæcum, but is especially so whenever it has been extravasated into the stomach; the acids of the gastric juice act upon the effused blood, so that it becomes black, and when discharged from the intestine it resembles a pitchy fluid, constituting true melæna.

The *symptoms* of diarrhœa have, perhaps, been sufficiently described in mentioning its several forms ; and they vary according to the cause. In the simplest form there is neither pain nor constitutional disturbance ; in more aggravated cases there may be severe colic, and febrile excitement; and generally, unless there be hepatic disturbance and derangement of the whole mucous tract, the tongue is clean, it is then furred and injected, and in typhoid prostration assumes a brownish color. The pulse is compressible, and the consequent prostration is often very alarming, especially in infants and aged persons, and in some cases it leads to a fatal result.

It is important carefully to mark the character of the evacuations; first, as to the admixture of undigested substances; secondly, as to the *fluidity* of the evacuations ; a simple fluid state, with normal excreta, indicates irritation of the mucous membrane in a slight degree; thirdly, the *presence of mucus* is evidence of more severe irritation of the colon; this is sometimes found in excessive quantity, and is easily recognized by pouring the evacuation from one vessel into another; fourthly, if more acute disease of the colon exist, detached portions of fæces are found floating on the fluid, which from the rapidity of its discharge, and possibly also from intestinal changes, is often frothy, from the admixture of air; fifthly, in severe diarrhœa, thin watery fluid may be discharged with scybala, and with sedimentary portions of fæcal matter ;[1] sixthly, thin fluid, almost like clear water, may be passed, as in some cases of albuminuria, from an œdematous condition of the membrane, or like rice-water in choleraic diarrhœa, or like soap-suds when with colliquative diarrhœa the hepatic secretion is also checked; seventhly, the fæces are sometimes discharged in a state indicative of fermentative action, and a frothy surface is produced of the appearance of yeast, and the whole discharge closely resembles the matters occasionally ejected from the stomach in obstructive disease at the pylorus; eighthly, as to the *color* of the evacuation, we have evidence thereby of the excess and of the paucity of *bile*, sometimes the stool being of a deep brown color, at others almost as pale as chalk; ninthly, the color may be changed by the admixture of such substances as logwood administered medicinally, or blackened by steel medicines, the sulphide of iron having been formed; and tenthly, the color is a guide to the detection of blood. Blood in the alvine discharges may be only observable by microscopical examination; but if in larger quantity, the color varies from the ordinary appearance of blood to the black pitchy stool of melæna, as we have before mentioned, according to the position of the hæmorrhage in the canal. The green color of the discharges in the severe diarrhœa of children, we believe, with Dr. Golding Bird, to be altered blood from an irritated and perhaps aphthous surface. Again, in severe dysentery, thin watery fluid, like the washing of beef, is sometimes discharged, consisting of blood with mucus, and of imperfect epithelial elements. To these dysenteric evacuations we shall have again

---

[1] Dr. Osborne "On the Examination of the Fæces," 'Dublin Quart.,' 1853.

to refer. Lastly, the *odor* of the fæces is not altogether unimportant; sometimes they are tolerably fetid from rapid degenerative changes, at other times they have scarcely any odor. In many instances the microscope enables us to detect an excess of mucus, the presence of blood, the rapid discharge of epithelial elements and nuclei, and other organic and inorganic substances, which the unassisted eye would in vain search for. We have elsewhere referred to the occasional presence of phosphatic crystals upon the mucous membrane of the intestines, and they are sometimes found in the alvine discharges, in simple as well as in typhoid diarrhœa. The presence of fatty matters in the evacuations was first noticed by Dr. Bright, in connection with disease of the pancreas; and the observations more recently made in reference to the physiological effects of the pancreatic fluid have directed increased attention to the subject. It must not be forgotten that we sometimes find oleaginous substances discharged after the administration of large quantities of milk and of cod-liver oil; thus in one case masses of fat as large as filberts were sent to me by a patient affected with phthisis, who had partaken of milk very freely; still, the observation has been confirmed by subsequent observers, that fatty matters are sometimes discharged in the alvine evacuations in disease of the pancreas, and sometimes in extensive disease of the mesenteric glands.

The *causes* of diarrhœa have been partially referred to.

(*a*) The most common cause of ordinary diarrhœa is exposure to cold and wet; standing in damp places; allowing the legs and loins to become damped and chilled; sitting down upon the ground, and falling asleep in the open air; injudicious bathing; the habit of leaving off flannel garments in hot weather, by which perspiration more rapidly evaporates, and the blood is driven from the surface towards the internal organs.

(*b*) Improper and indigestible food, unripe fruit, and an *excess* of uncooked fruit; salads, pastries, and much that modern cookery produces, especially when an excess in quantity is combined with an injurious quality.

In infants a fertile source of diarrhœa, often passing into severe gastro-enteritis, is the administration of unsuitable food, the injurious effects of which are greatly increased by exposure to cold. In hospital and dispensary practice this cause of disease is observed to a frightful extent; at seven or eight months, even while the infant is, in a great measure, nourished by the breast of the mother, meat, raw vegetables, and fruits, sweets, almost *ad libitum*, are given; and a few months later we often find, that before the child has the power of mastication the mother gives the food of which she herself partakes, sometimes adding malt liquors and ardent spirits. The consequences of this dietary are such as might be anticipated; the food passes onwards undigested, severe gastro-enteritis is induced; and the malady is often aggravated by a want of cleanliness, and by exposure to night air and dampness. The mortality in London from these causes is exceedingly great. In other infants the food, although in itself proper, is unsuited to the condition then existing, and perpetuates diarrhœa; or it may be that the milk of the mother disagrees with the child, from the impairment of her health. In such subjects we occasionally find, that an alteration in the character of the gastric juice of the infant leads to coagulation of the milk, and to severe diarrhœa, with colic, etc., the stools containing portions of curdled and undigested milk, namely, oleaginous matter mixed with casein.

(*c*) Diarrhœa is set up by exhaustion, either from want of food, starvation and its attendants of misery, or as the consequence of chronic disease. This form of diarrhœa is sometimes observed in women who have nursed their infants too long. Enfeebled by bearing children rapidly, their strength is additionally taxed by nursing for twelve, fifteen, or eighteen months without proper nourishment or invigorating air. The whole mucous membrane is affected; the nerve of organic life shows its ebbing powers; the blanched cheek, the dilated pupil, .the desponding countenance, and impulses of a mind verging on insanity, are symptomatic of this condition. There is intense pain in the head, the heart is enfeebled, the pulse sharp, and sometimes irregular; there is a distressing sensation of exhaustion at the scrobiculus cordis, with severe pain in the back, and in this state a very slight irregularity of food will sometimes set up diarrhœa and vomiting. Cancerous and strumous disease of the mesenteric glands, obstruction of the thoracic duct, chronic disease of the pancreas, diabetes, etc., sometimes have uncontrollable diarrhœa as one of their latest symptoms.

(*d*) *Epidemic causes.*—At some seasons of the year, in our own climate during the spring and autumn months, diarrhœa of varying severity is set up, and appears to arise from the condition of the atmosphere, perhaps from germs of vegetable or animal growth.

(*e*) *Endemic causes* are more numerous, and with them may be classed the diarrhœa arising from offensive drains, from decaying animal and vegetable matters. Causes of this kind operate with greater severity upon the young and enfeebled, upon the strumous and ill-nourished. Many infants are thus affected with diarrhœa, and with severe general gastro-enteritis. It is now well known that an impure water-supply, especially if contaminated by sewage, will lead to diarrhœa as well as to enteric fever, and probably to cholera. Again, a general dampness of locality, as from a clay subsoil, will set up, or will increase and perpetuate diarrhœa. We have witnessed the removal into dry bracing air followed by cessation of the disease, and the return to the same district repeatedly cause its recurrence.

(*f*) *Excessive secretion* of bile, and other diseases of the liver, as well as disease of other intestinal glands, set up diarrhœa.

(*g*) Other causes are, tubercular disease of the mucous membrane of the intestine and the mesenteric glands; œdema and long-continued congestion of the mucous membrane; mental agitation and fright; ulceration of the small and large intestine, as in fever, phthisis, &c. ; cancerous diseases; purpura and scurvy; large draughts of water; miasmatic disease; poisons.

Prognosis.—Diarrhœa is never altogether free from danger in aged persons, or in very young children; but the prognosis differs according to its cause and character. If associated with chronic disease, or an enfeebled condition of the system, it is often the immediate precursor of death; but when the cause can be removed, and the subject is young, however severe the case may be, we should encourage the prospect of recovery. Many of such cases, when apparently quite *in extremis*, have gradually and almost miraculously recovered.

The prognosis is unfavorable, when diarrhœa has been long-continued, and is very severe in its character; in some of these cases scarcely any treatment appears to arrest the purging, and the patient gradually sinks into a typhoid condition.

It may appear unnecessary to say anything in reference to the *diagnosis* of diarrhœa; it is well, always, if possible, to ascertain personally the character of the evacuations; since there may be apparent diarrhœa, without the reality. I have seen starch enemata used when patients were greatly exhausted, and, on inspection, found the intestine loaded with solid fæcal matter. In spinal disease, a weak sphincter ani with involuntary defecation is often mistaken for diarrhœa, and I have known astringents continued for several months ineffectively, whereas rest to the spine quickly relieved the malady. A hardened mass of fæces, which the patient is unable to expel from the rectum, frequently leads to the repeated evacuation of small quantities of fluid fæces or of mucus, which is regarded as diarrhœa or even dysentery; the effort at expulsion is constant and painful, but ineffective; the removal of the mass at once checks the supposed diarrhœa. Or again, in an exhausted state of the system, or during epidemic diarrhœa, a single loose motion may require immediate attention; for the character rather than the quantity should be our guide. In persistent diarrhœa it is important always to examine the rectum, for I have frequently known cancerous disease entirely overlooked from the want of digital examination.

Treatment.—The primary object must be to ascertain the character of the diarrhœa, and to remove, if possible, its cause. If food be improper, to change it, and administer such as shall be of the least irritating kind. If the air be impure, to order removal to a healthy atmosphere. If the mucous membrane and the secretions be disordered, to try and restore them to a healthy state. To check the diarrhœa by various astringents and by rest.

(*a*) *Warmth.*—Warm baths, warmth applied to the feet, and flannel to the abdominal parietes, a warm but pure air, &c., assist in checking many of the simpler forms, and in diminishing those arising from chronic disease. Local warmth may be attained by the application of a hot fomentation, or poultice to the abdomen, or by such rubefacients as a mustard poultice, or turpentine embrocation.

(*b*) *Food.*—In diarrhœa the least irritating and the most easily digestible kinds of nourishment are advisable. Many of the forms of amylaceous aliment, arrowroot, sago, are of this kind, and may be given made with milk; these are in themselves soothing applications to irritated mucous membranes, whilst they serve as nourishment to the system. Milk, rice, soaked bread and toast, lightly-boiled puddings of flour and eggs, &c., may be also taken with advantage, and in chronic diarrhœa suet and milk is often of great benefit.

(*c*) The avoidance of *stimulants*, of rich and greasy food, of highly seasoned dishes, of vegetables, especially when uncooked, of fruits, &c., is essential; and it is well in many cases to abstain for a short time from solid animal food altogether. The forms of animal food which are most easily digestible are chicken, sweetbread, and some forms of fish, as sole, cod, and whiting; then venison, mutton, and beef; but much depends on the mode in which these viands are dressed. When dried, salted, and cold, they require a much longer period for their digestion, and portions often pass into the intestine undissolved. Beef-tea sometimes appears to increase diarrhœa, when veal and mutton-broth can be taken with benefit.

(*d*) *Rest*, and the avoidance of muscular excitement and sudden movements, are very important in checking diarrhœa; and in many instances, especially in severe cases, a recumbent posture should be maintained. In the erect position the gravitation of fluids increases their rapid movement over the irritated mucous membrane.

(*e*) *Pure and dry air* is very desirable; many patients at once recover when removed from a damp atmosphere to a dry and bracing one; and when the contamination of decomposing animal and vegetable substances is setting up the disease, removal is still more important, and is often essential to permanent restoration. In miasmatic districts, diarrhœa may not only be rendered paroxysmal, but be perpetuated by the marsh poison.

Many cases of diarrhœa will be cured by this attention to warmth and diet, to rest and pure air; but other means often promote the comfort and favor the restoration to health.

If the large intestine, and especially the rectum, be affected, much benefit is derived from enemata. These are composed of various ingredients, simple starch, thin gruel, and barley-water; and to these we may add tincture of opium and biborate of soda. Or they may be made as tringent, as decoction of oak bark with tragacanth, or glycerine of tannin with water; or a very dilute solution of nitrate of silver may be used; an infusion of ipecacuanha has been favorably recommended as an injection by Boudin and Chouppe.

To restore the diseased mucous membrane and to correct secretions.—(a) The alkalies are of very great service in diminishing congestion, as well as in rendering the secretions less irritating. Solution of potash, limewater, chalk, some salines, as chlorate of potash, bicarbonate of potash, and nitrate of bismuth, act in this manner.

(b) When the hepatic secretions are disordered, as shown by furred tongue, and pale evacuations, the moderate use of Mercurials is of value, as gray powder or calomel, combined with Dover's powder, with soda or with opium; but we should strongly urge that mercurials be very carefully administered, because in many forms of diarrhœa they tend greatly to aggravate the disease. It is only in some cases, even with a foul tongue and deficient hepatic secretions, that we would recommend their use.

(c) *Demulcents.*—These act by directly sheathing the mucous membrane; the most important are those mentioned as food, but others are of considerable utility, as acacia, tragacanth, linseed, liquorice, glycerine, spermaceti, &c.

(d) *Castor-oil, Linseed-oil.*—These are of great value, when improper food, retained secretions and scybala irritate the alimentary canal. They are combined with great advantage with the compound tincture of rhubarb, and sometimes with a small dose of opium, ℈v. or x. These remedies are of most service in some forms of dysenteric diarrhœa, when scybala irritate the mucous membrane.

(e) *Ipecacuanha* is a remedy which acts, apparently, on all the mucous membranes, and is as valuable in disease of the alimentary as of the respiratory mucous membrane. Ipecacuanha not only increases the quantity of mucus but it mitigates inflammatory congestion. It is of great service in the dysenteric diarrhœa of adults, and equally so in the diarrhœa of infants. In the former, Dover's powder is a valuable form for its administration, or the ipecacuanha may be combined with astringents, as in the compound infusion of krameria,[1] and the compound logwood mixture of the Guy's Pharmacopœia, or it may be administered alone, as in the treatment of pure dysentery.

(f) *Astringents and Desiccants.*—These may be divided into several classes. The *saline*, as chalk; the *vegetable*, as tannic and gallic acids, krameria, kino, catechu, logwood, Indian bael, cusparia, opium; *metallic*, as sulphate of copper, acetate of lead, nitrate of silver, nitrate of bismuth, &c.

---

[1] Infusum Krameriæ compositum. Infusion of Rhatany Root, fl. ℥ xj. ; Ipecacuanha Wine, ℨ iv., Tincture of Catechu, ℨ iv.

(*g*) *Opium* acts not only as an astringent, but also as a narcotic; it diminishes the secretion from the mucous membrane, and the peristaltic movement of the intestine, and it relieves the pain of colic. It is of great value in diarrhœa, and may be combined with other remedies, as with chalk and ipecacuanha; but, when irritating ingesta and disordered secretions perpetuate diarrhœa, opium and astringents are not appropriate remedies. When the disease is chronic, opium may be given with the more active vegetable astringents, catechu, krameria, and logwood, and sometimes very advantageously with quinine.

The metallic astringents are combined in a similar manner with opium and ipecacuanha, but are more frequently used in chronic dysentery, and in tubercular ulceration of the intestine, than in simple diarrhœa.

(*h*) *Mineral Acids.*—Much has been written upon the use of dilute sulphuric acid in diarrhœa ; and its use has certainly been attended with benefit, although not to the extent we were led to suppose. Both dilute sulphuric acid, and dilute nitric acid, are of value after the more severe symptoms have passed off ; they act at first possibly by checking chemical and fermentative changes, and afterwards as tonics to the relaxed mucous membrane. Combined with slightly astringent and mucilaginous tonics, as with cusparia and simaruba, or with calumba root and elm bark, they are of great service in some cases.

When there is much pain we may associate narcotics with other remedies before mentioned. Spirit of chloroform and spirit of camphor in small doses sometimes afford great relief, so also the tincture of henbane; in other cases simple carminative medicines are sufficient to relieve the pain, as ginger, cardamoms, &c., especially where the diarrhœa is associated with flatulent colic.

In the colliquative diarrhœa of weaned children Dr. I. F. Weisse has strongly advocated the administration of raw meat, scraped and reduced to a pulp, as we have previously mentioned in the remarks on enteritis.

(*i*) *Leeches.*—The application of leeches to the anus is a remedy which greatly relieves inflammatory congestion of the mucous membrane of the large intestine, but it is one which we should scarcely recommend, unless the disease assume a severe and dysenteric character.

(*j*) *Suppositories*, composed of the compound soap pill or morphia, are often of great service when there is distressing tenesmus which disturbs the rest of the patient; and when it is undesirable to administer an opiate by the mouth, or inconvenient to use an enema. Tannin may also in this way be conveniently used, so also bismuth.

In chronic mucous discharge from the bowels, we must first seek to remove the disease of the liver, if such exist, by mild alteratives, by taraxacum, and by nitro-muriatic acid. These remedies, also, assist in relieving the chronic congestion and inflammation of the intestine, and are more effective than astringents. It is well, however, to be assured that no polypoid growth, nor disease of the rectum and sigmoid flexure, is setting up the disease.

If astringents be required in these instances, the oxide and nitrate of silver, sulphate of copper with opium, or the vegetable astringents just mentioned, may be used; and as enemata, glycerine of tannin diluted with water, the solution of nitrate of silver (gr. x.—xv. to Oj.¹), the infusion of quassia, the decoction of oak bark, and the decoction of poppies with or without the addition of borax, may be employed with advantage.

---

¹ Trousseau.

In the treatment of choleraic diarrhœa, rest in the recumbent position, warmth to the abdomen and the feet, and gentle friction, if muscular spasm distress the patient, are valuable remedies. A full dose of chalk and opium with catechu and aromatic spirit of ammonia should be given, and repeated in two or three hours, if necessary. Demulcent nutriment, as mutton-broth and arrowroot, may be allowed; and if vomiting supervene, ice or cold water will be beneficial. Dilute sulphuric acid has been sometimes used with great advantage, and by some calomel has been freely given in these cases, especially when vomiting has come on. When the collapse of true cholera has attacked the patient, general experience does not favor the free use of either opium or brandy; but to enter fully into the treatment of cholera is foreign to our purpose.

The following cases of diarrhœa are of considerable interest:

CASE CXXV.—*Inanition. Diarrhœa.*—John M——, æt. 26, was admitted into Guy's Hospital Dec. 17th, 1856, and died Dec. 20th. He had been a sailor, and stated that he had had dysentery, but this was not satisfactorily ascertained, on account of his prostrate condition. It appeared that he had been on board an American vessel from China to Liverpool, and arrived at the latter place on December 6th; he then came up to London. He informed the nurse that there had been a mutiny on board, and that he had been put in irons in the hold. He was in the most emaciated state; the voice was scarcely perceptible; the pulse was exceedingly compressible, and the tongue and mouth presented yellowish white aphthous patches; he had no vomiting, but the stools escaped from him, and were white and very offensive; the respiration was easy, and the mind perfectly conscious. Milk was ordered. The following day he was better, but sank on the third day after admission, and was sensible til. nearly the last.

Inspection, December 22, 1856.—There were ecchymoses on both thighs, and old cicatrices on the wrist and leg. The brain was less firm than normal; the lungs were collapsed and healthy. The heart was small. The liver was healthy. The gall-bladder was not distended, and the spleen and kidneys were healthy. The stomach presented gastric solution at the cardiac portion. The small intestines were healthy. The large intestine was throughout of a gray color, and was filled with dry, white fæces. At the root of the mesentery were several white strumous masses in the glands, but it could not be found that the thoracic duct was obstructed. The urinary bladder was distended.

This case presents us with a well-marked instance of a man dying from the effect of starvation. The diarrhœa was probably the result of want of nourishment, of good air, and of light, &c.; so that supplies having been cut off and the conditions necessary for reparation excluded, the whole body wasted, and the spark of life gradually expired.

CASE CXXVI.—*Chronic Diarrhœa. Hysteria. Great Relief from Tincture of Iron.*—Georgiana B——, æt. 40, a single woman, who had resided in the Commercial Road, and had supported herself by her needle, applied at Guy's Hospital May 23, 1860, and was admitted under my care. She had suffered from uterine ulceration. During eighteen months she had been affected with diarrhœa, and when she had mental anxiety the disease increased in severity. The slightest exertion produced perspiration. She was a tall woman, extremely nervous, the eyes sunken, the countenance dejected. The heart and lungs were normal. She complained of great pain in the abdomen, on the right side below the liver, in the region of the ascending colon; there was tympanitic distention in the same region; the bowels were opened six times in twenty-four hours, but there was no evidence that blood had been passed. She had not taken meat during several months. Astringents of different kinds were administered and enemata used, with only partial relief, till the tincture of iron was given persistently for several weeks. The diarrhœa then subsided, and she left the hospital convalescent, stating that she had not been so well for eight years.

This case appeared to be one of passive mucous diarrhœa in a very hysterical subject, and the uterine irritation had tended to perpetuate the disease. Astringents were less efficacious than preparations of steel, which diminished the nervous irritability and gave tone and strength to the whole system. The regulated and more generous diet which was given must not be overlooked; and by persuasive measures she was induced to take a meat diet, which lessened the fluid contents of the colon, and thereby increased the consistence of the alvine discharges.

DYSENTERY.

By J. Warburton Begbie, M.D., F.R.C.P.E.

Definition.—A febrile disease, in which inflammation affecting the glandular structures of the large intestine chiefly—although sometimes extending to the small—and producing ulceration, tends to terminate in sloughing of the mucous membrane. The disease is accompanied by much nervous depression, and is characterized by tormina—severe pains in the abdomen of a griping nature—followed by frequent scanty and bloody stools, straining, and tenesmus.

The term Dysentery is derived from two Greek words—δυς, hard or bad, and ἔντερον, a piece of the guts, intestines. Δυσεντερία, was itself employed by Hippocrates and other Greek writers to signify a bowel complaint, or bloody flux.

Synonyms.—Tormina; Tormina intestinorum; Fluxus dysentericus; Fluxus cruentus; Fluxus torminosus; Rheuma ventris; Febris dysenterica; Colonitis; Bloody Flux; Dysenteria; Flux de Sang (French); Die Ruhe, Die rothe Ruhe (German); Dissenteria (Italian); Dysenteria (Spanish).

History.—Dysentery has been known as a disease since the earliest period of medical history. In several of the Hippocratic treatises, but especially in the following,—Περὶ ἀέρων, ὑδάτων, καὶ τόπων, Προγνωστικόν, and Ἀφορισμοί,—are many interesting remarks regarding the symptoms and treatment of Dysentery, also the prognosis to be founded upon it, to be met with. Aretæus has described Dysentery with his usual conciseness, and even more than his usual ability. In Cælius Aurelianus, but still more in Celsus, much information may be found regarding Dysentery, as the disease was known in the days of these celebrated Latin writers. Coming down to modern times, Sydenham, Ramazzini, Morton, Huxham, Cleghorn, Morgagni, Zimmerman, and Sir John Pringle (in his celebrated treatise on Diseases of the Army), are among the more distinguished of the numerous writers on Dysentery.[1]

Dysentery is placed by Cullen in class first, "Pyrexiæ," and of it the fifth order, "Profluvia." Of the latter his definition is "Pyrexia cum excretione aucta naturaliter, non sanguinea." Dysentery, Cullen defines as follows: "Pyrexia contagiosa; dejectiones frequentes, mucosæ, vel sanguinolentæ, retentis plerumque fæcibus alvinis; tormina; tenesmus."[2]

It is customary to distinguish between acute and chronic Dysentery, also between epidemic and non-epidemic or sporadic Dysentery. To the

---

[1] For a full and instructive account of the history and geographical distribution of Dysentery, see Hirsch, "Handbuch der historisch geographischen Pathologie," article "Ruhe," vol. ii. p. 194.

[2] Synopsis Nosologie Methodicæ, p. 308.

non-epidemic disease we are now to direct attention—the epidemic Dysentery having been already considered by Dr. Maclean.

SYMPTOMATOLOGY.—The essential characters of Dysentery are severe pains of a griping nature in the belly, followed by frequent and bloody stools, defecation being accompanied by much straining and tenesmus. The later symptoms are the most characteristic. Watch a patient affected by Dysentery at stool: he sits a long time, straining; his features are distorted by the pain he suffers; the discharge from the bowels may be, often indeed is, but scanty: still he sits. The strong desire to remain at stool, accompanied by griping and straining, is expressed in the word tenesmus. Scarcely can such patients at times be persuaded to leave the stool and return to bed, until they feel so faint as to be unable longer to maintain the sitting posture, and sometimes while on the stool they faint.

Straining and tenesmus do not occur in diarrhœa, they are peculiar to dysentery; and so also are the other symptoms, named in Cullen's definition; the passage of blood and mucus, the fæces being for the most part retained, or after a time passed in the form of small, often hard, scybala.

Acute Dysentery.—The disease in this form may occur without any premonitory symptoms; more commonly, however, it is preceded by such. General uneasiness, lassitude, impaired appetite, disagreeable sensations in the abdomen, confined bowels, or a loose condition of the bowels, are among the more frequent of the premonitory symptoms. These may have existed for a few days, when a chill is experienced, or sometimes a chill or rigor is the very earliest indication of departure from a healthy state. To these succeed the febrile symptoms, heat of skin and quickness of the pulse. Much variety exists in respect to the degree of the general or constitutional disturbance which accompanies the local affection in Dysentery. That may be very slight indeed; the disease may even run its course without fever. On the other hand, the constitutional disorder may be severe, and is not unfrequently profound, assuming an adynamic or typhoid character. In the simpler variety of the disease, there are at the commencement griping pains in the belly, those pains to which the name of "tormina" is now generally applied. This term was first used by Celsus.[1] "Proxima," he says, "his inter intestinorum mala *tormina* esse consueverunt: δυσεντερία Græcè vocatur." The tormina are felt in different parts of the belly, and, like the pain of colic, yield at one time, to return again, perhaps more severely than before. With the tormina there occur discharges, usually slight, from the bowels, and by these a partial relief to the pain is experienced. To the tormina and diarrhœa succeeds the tenesmus; and this term may be understood as including the frequent desire to go to stool, and the reluctance to leave it, with the very distressing feeling of bearing down, and burning sensation in the rectum. In every marked case of Dysentery the tenesmus is a prominent as it is the most distressing symptom. The discharge from the bowels

---

[1] Liber iv. ch. xv. The description of the disease given by Celsus is so accurate as to merit perusal; the earlier sentences may be quoted. "Intus intestina exulcerantur; ex his cruor manat, isque modo cum stercore aliquo semper liquido, modo cum quibusdam quasi mucosis excernitur; interdum simul quædam carnosa descendunt; frequens dejiciendi cupiditas, dolorque in ano est; cum eodem dolore exiguum aliquid emittitur: atque eo quoque tormentum indenditur; itque post tempus aliquod levatur; exigua requies est; somnus interpellatur; febricula oritur; longoque tempore id malum, cum inveteraverit aut, tollit hominem aut, etiamsi finitur, excruciat."

affords little relief when the tenesmus is great. The calls to stool of course vary greatly in frequency: in some instances they are almost incessant. Occurring in children, particularly, but occasionally also in adults, as a consequence of the frequent evacuations, and the tenesmus by which they are accompanied, is prolapsus of the anus, which in itself requires careful management, and may become a very troublesome sequela of the disease.[1]

The discharges from the bowels in Dysentery are peculiar and characteristic. At first they are usually feculent, if not entirely, at least chiefly so; but very soon, becoming very scanty in amount, they are found to be composed of mucus, or of mucus mixed with blood, and sometimes of nearly pure blood. When the inflammation of the bowels·has advanced to a certain stage, it is common to notice the appearance of vitiated bile in the stools, and likewise of shreddy-looking portions of fibrine or false membrane. The odor of the evacuations in Dysentery is one *sui generis*, quite peculiar, and highly offensive. Not unfrequently there is sympathetic irritation of the bladder, and a frequent as well as difficult micturition. While the chief part of the pain in Dysentery is experienced during the movement of the bowels, it is not limited to that time—pain is present in the abdomen generally aggravated by pressure. When, in addition to the tenderness over the left side of the belly, corresponding to the position of the sigmoid flexure, there is pain felt over the epigastrium and down the right side, it may be conjectured that the disease has implicated the large intestine in its entire extent, and is not limited, as happens in milder instances, to the rectum and descending portion of the colon.

More or less of fever accompanies Dysentery. In mild cases the feverish disturbance, as already stated, is slight, but, on the other hand, in the more decided instances of the disease, the constitutional disturbance is evidenced by the quickness of the pulse, the augmented temperature of the surface scanty secretion of urine, and the coated condition of the tongue. In the milder cases of Dysentery there is no special implication of the nervous system; the pulse in such, although frequent, is full and of good strength: neither nausea nor vomiting, except of occasional occurrence, are present; and although the local malady may be severe, the disease wears throughout a sthenic character. But it is not always so; an asthenic or adynamic form of Dysentery also occurs, characterized by a frequent, small, and feeble pulse, pallor and coolness, rather than warmth of the skin, the occurrence of a clammy moisture over it, anxious expression of the countenance, sunken eyes, dryness and glazing of the tongue, suppression of the voice, hiccough, delirium, prominence of the abdomen and rapid sinking. With these indications there is unusual violence in the local symptoms, particularly as regards the frequency of the discharges from the bowels. These ultimately become exceedingly offensive and watery. They present the appearance of water in which raw flesh has been washed, and are known by the name of "lotura carnium." The disease may thus prove fatal in a few days. Dr. Wood speaks of such cases as very rare, and only seen during epidemics.[2] The latter observation is no doubt correct, but only to a certain extent, for these instances of rap-

---

[1] "Durch die heftigen Anstrengungen wird auch nicht selten, zumal bei Kindern, ein Prolapsus ani herbeigeführt der sich entweder von selbst weider zurückzieht oder reponirt werden muss."—HENOCH, *Klinik der Unterleibs-Krankheiten*, Ruhe, Band 3, p. 235.

[2] A Treatise on the Practice of Medicine. By George B. Wood, M.D. Vol. i. p. 625.

7

idly fatal dysentery, although more common in the epidemic prevalence of the disease, are occasionally met with in the non-epidemic malady. It has occurred to the writer to witness one or two very rapidly fatal cases of Dysentery, in which a remarkable depression of the nervous system was evident from the very commencement of the disease. In the ordinary form Dysentery tends to a favorable termination, and usually before the lapse of a week or eight days there are indications of amendment. The acute disease sometimes terminates in chronic dysentery.

Chronic Dysentery is characterized by the frequency of the evacuations, which, at the same time, are usually very scanty. As in the acute affection, so in the chronic, the discharges are attended by local suffering and tenesmus. Mucus, or mucus mixed with blood, sometimes with purulent matter, constitutes the bulk of the evacuations; feculent stools occur when the disease, instead of implicating the entire colon, is limited to the rectum, or involves with it only the descending portion of the former. Chronic dysentery may last for months or years. In some instances it appears to produce wonderfully little influence on the general health and strength of the invalid, but as a general rule the sufferer from chronic dysentery is emaciated, pale, and weakly; and the disease is not unapt to prove fatal, through the exhaustion consequent upon its long continuance, or owing to the establishment of a state of continual or hectic fever.

Among the morbid conditions which are connected with, or result from, attacks of Dysentery, whether acute or chronic, affections of the liver occupy a chief place, and to these attention will be called in treating of the pathological anatomy of the disease. Anæmia, more or less marked, results from Dysentery. The writer remembers to have seen a case of anæmia of a very typical character, in which the blood impoverishment was due to a long-continued attack of Dysentery. To the occurrence of paralysis in conjunction with Dysentery, Romberg has called attention,[1] and he quotes a passage from an old dissertation by Fabricius: " De paralysi brachii unius et pedis alterius lateris dysentericis familiari,"[2] in verification of the remark. J. P. Frank refers to the same occurrence;[3] and although Graves[4] has not specially mentioned Dysentery as a form of intestinal disease giving rise to a reflex paraplegia, he has emphatically done so in reference to Enteritis. By Zimmerman[5] and Joseph Frank[6] allusion is made to paralysis of the arms and legs occurring after Dysentery.

MORBID ANATOMY.—As Dysentery is essentially a disease of the large intestines it is in the colon and rectum that we look for the morbid appearances characteristic of its occurrence.[7] The mucous membrane in these

---

[1] " Auch bei der Dysenterie," remarks Romberg, " ist das Vorkommen der Paralyse beobachtet worden."—*Lehrbuch der Nervenkrankheiten des Menschen : Spinale Lähmungen.*

[2] Disputationes ad Morborum Historiam et Curationem facientes quas collegit edidit, et recensuit Albertus Haller. Tomus primus, p. 97.

[3] " Tantum verò ad gradum doloris in abdomine vehementia apud hos vel illos evehitur, ut ab eo non minus ac in colica saturnina brachii aut pedis unius vel alterius paralysis sequatur."—*De Curandis Hominum Morbis.* Auctore Joanne Petro Frank, Liber v. De Profluviis, pars ii. p. 497.

[4] Clinical Lectures, Edition 1864, in one vol. p. 415.

[5] Von der Ruhe unter kem Volke in Jahr 1765.

[6] Praxios Medicæ Universæ Præcepta. Auctore Josepho Frank. De Paralysi.

[7] The mucous membrane of the colon, says Rokitansky, is the seat of the dysenteric process ; and we may state it as a rule, that its intensity increases from the cæcal valve downwards, and consequently is met with in the most fully-developed state in the sigmoid flexure and in the rectum. It not unfrequently passes beyond the cæcal valve towards the ileum, but is here only seen in its mildest form.

portions either presents the appearance of having been diffusely inflamed, being everywhere much reddened, thickened, and at parts ulcerated, or, with the absence of diffuse inflammation, there exists remarkable prominence of the solitary glands and mucous follicles. There exist three separate and distinct forms of ulceration affecting the mucous surface of the intestines—the tubercular, the typhoid, that met with in enteric fever, and the dysenteric. Apart from other characteristic differences in these affections, the last-mentioned is nearly limited in its occurrence to the large bowel, while the two former are especially met with in the small intestines, and particularly in the ileum. The size of dysenteric ulcers varies. They are sometimes small, and present a nearly circular form, or they are larger, irregular in shape, having an abrupt border, are covered by a·dark-colored slough, and appear as if formed by the coalescence of several smaller ulcers. It is not uncommon to find considerable portions of more or less dense lymph, coating the reddened and thickened mucous surface. Portions of false membrane having precisely the same appearance are sometimes passed at stool; but these, while still adherent to the bowel, do not when removed usually disclose an ulcerated surface. A truly sphacelated condition of the mucous membrane is occasionally met with, and pieces of gangrenous mucous membrane, sometimes of considerable size, have been passed in the evacuations in certain cases of Dysentery. Perforation of the bowel, which is of no uncommon occurrence in the progress of typhoid ulceration, and occasionally takes place in tubercular disease of the bowels, is very rarely indeed met with in Dysentery: the mucous, sub-mucous, and muscular coats of the colon suffer in this disease, but the peritoneal covering is not so apt to be involved. The mesenteric glands in Dysentery are frequently found tumefied and presenting a dark-bluish color. They may be softened, but are very rarely indeed the seat of suppuration. Even when much enlarged, they have not been distinguished by the presence of any peculiar morbid product such as occurs in the typhoid and tubercular tumefactions of these glands. Rokitansky describes the dysenteric process as divisible into four natural degrees or forms.[1] The anatomical characters of the *first* or lowest form are, swelling, injection, and reddening, softening (red and bleeding), serous exudation in the shape of a delicate vesicular eruption, and consequent branny desquamation of the epithelium (the latter appearance probably led Linnæus to term Dysentery " Scabies intestinorum interna "). In the *second* form, a larger surface of the bowel is involved, but still presenting a deeper development at one part than another—there is copious infiltration of the sub-mucous cellular tissue, giving rise to a greater or less number of prominences, which correspond to those parts of the mucous membrane at which the morbid process is most conspicuous. The intestine is generally in a state of passive dilatation, distended by gas, and occupied by a dirty-brown fluid, composed of intestinal secretions, epithelium, lymph, blood, and fæces. The coats of the bowel are thickened, and the sub-mucous· tissue especially in a state of tumefaction. In the *third* stage, the prominences are more thickly set, and the result is an uneven lobulated appearance. The mucous membrane investing these prominences is in part converted into a slough, or it may have disappeared, so as to expose the infiltrated sub-mucous cellular tissue to which the remnants of the mucous membrane remain attached, in the shape of solitary dark-red, flaccid, and bleeding vascular tufts, or as dilated follicles which are capable of easy

---

[1] A Manual of Pathological Anatomy. Sydenham Society's Edition, vol. ii. p. 83.

removal. The contents of the intestine are now of a dirty-brown or red dish, ichorous, fetid, flocculent and grumous character. In the *fourth* and highest degree, the mucous membrane has degenerated into a black, friable, carbonified mass, portions of which may be subsequently voided in the shape of tubular laminæ (so-called mortification of the mucous membrane). The sub-mucous cellular tissue appears to be infiltrated with sero-sanguinolent fluid, or dark blood; or it is pale, and the blood contained in its vessels is converted into a black solid mass. Purulent infiltration of the sub-mucous tissue is also found. The affected portion of the bowel, which contains a putrid fluid resembling coffee grounds, may be either in a state of passive dilatation, or (and this is more frequently the case) collapsed. In the higher degrees of the dysenteric process the muscular coat of the colon suffers; its tissue becomes condensed, pale, ashy, and friable. In the same degrees, the peritoneal covering does not completely escape, it presents a dirty-gray discoloration, has lost its lustre, and here and there dilatation and injection of its capillary vessels is visible, while occasionally it is covered by a thin brownish ichorous exudation. These characters afford the means of recognizing the existence of an advanced stage of Dysentery, while as yet the intestine has been unopened, and the mucous surface unexposed. Rokitansky has some very interesting observations on the termination of Dysentery. Provided disorganization of the mucous membrane has not occurred, a cure results through the return of normal cohesion, and the generation of a new layer under the desquamated epithelium. In the more intense degrees of the dysenteric process, and when disorganization has taken place, the mucous membrane having undergone more or less destruction, one or two results ensues— either a real cure of the loss of substance, with consolidation of the abraded portions of the intestine, follows, or the entire process assumes a low chronic form, the specific nature of the disease is lost, and an inflammation atonic in character, with suppuration of the intestinal coat, occurs. Dysentery is fatal through the more or less rapid, or more or less penetrating, destruction of tissue and coincident exhaustion. When cure results, the loss of substance having been inconsiderable, new tissue is formed, and may so contract as to bring the edges of the mucous membrane into apposition with one another, while a cicatrix remains, which has the appearance of a large number of agminated warty excrescences of the mucous membrane between which the sero-fibrous basis from which they proceed may be detected. On the other hand, in those instances of the disease which have been distinguished by an extensive loss of substance, the approach of the edges is impossible, and the deeper layers of the tissue which takes the place of the mucous membrane are frequently condensed into fibrous bands, which form projections into the intestinal cavity, interlaced with one another, and not unfrequently encroach upon the calibre of the intestine, in the form of valvular or annular folds, thus giving rise to a variety of stricture of the colon.

Reference has already been made to the participation by the liver in disease in connection with Dysentery. Abscess of the liver has been supposed by some authorities to have an intimate relationship to the dysenteric process in the colon. Of the not unfrequent association of the two diseases there can at all events be no question. Dr. Parkes[1] found, in twenty-five cases of Dysentery, seven to be affected with hepatic ab-

---

[1] Remarks on the Dysentery and Hepatitis of India, 1846.

scess. In the large work of Mr. Annesley,[1] there are twenty-nine cases of abscess of the liver recorded, and of these no fewer than twenty-one, or nearly three-fourths, had ulceration, more or less extensive, in the large intestine, while in two other cases there were appearances of constriction and contraction which were reasonably ascribed to the existence of Dysentery at some former period. Annesley regarded the Dysentery as the result of the disease of the liver, or hepatitis. By certain writers, among whom Dr. Abercrombie[2] and the late Dr. William Thomson of Glasgow[3] may be mentioned, the concurrence of the two diseases has been regarded as accidental. The former observes: "Dysentery is often accompanied by diseases of neighboring organs, especially the liver, in which are to be found, in some cases abscesses, in others, where protracted in their duration, chronic induration. These are to be regarded as accidental combinations, though they may considerably modify the symptoms." A third view, and one which has been popular in this country since it was ably upheld by Dr. Budd,[4] is that the inflammation of the liver terminating in abscess is the result of purulent absorption from the dysenteric process in the colon. Many years ago, Andral and Louis, apparently unsuspecting any connection between hepatic abscess and ulcerated intestines, noticed the co-existence of the former with ulceration in the large intestines and in the lower end of the ileum in two cases, in the lower end of the ileum alone in one case, in the stomach in four cases, in the gall-bladder in one. In one of the cases in which the stomach was affected, Andral concludes with reason that the ulcer was caused by the hepatic abscess bursting into the stomach. But excluding this observation, there resulted seven out of fifteen instances of hepatic abscess, in which there existed at the same time ulceration in some part of the extensive mucous surface which returns its blood to the portal vein. These observations of the French pathologists were very far indeed from being singular. Thus Dr. Cheyne, of Dublin, in writing of the Dysentery in Ireland, remarks that in the majority of his dissections the liver was apparently normal, but that in two cases he found abscesses in its substance. But while the occasional intimate connection of hepatic abscess with Dysentery, and of which Dr. Budd's theory in all probability assigns the true cause, has been determined, it must also be admitted that abscess of the liver frequently occurs in tropical countries wholly unconnected with Dysentery, not acknowledging a pyæmic origin, and not resulting from mechanical injury. Dr. Murchison, of London, in his papers on the Climate and Diseases of Burmah,[5] pointed out that, in many cases, abscess of the liver met with in tropical countries occurred independently of these three causes. Dr. Morehead,[6] while admitting the occasional occurrence of hepatic abscess, according to Dr. Budd's explanations—that is, by the transmission to the liver of pus or vitiated secretion originating in an ulcerated intestinal surface,—is satisfied that, as a general proposition, such a view is altogether at variance with the results of clinical research in India. Seventeen cases of hepatic abscess are detailed by Dr.

[1] Researches into the Causes, Nature, and Treatment of the more prevalent Diseases of India and of Warm Climates generally. By James Annesley, 2 vols. 4to. London, 1828.
[2] Researches on the Pathology of the Intestinal Canal. 1820.
[3] Practical Treatise on the Diseases of the Liver and Biliary Passages. 1841.
[4] On Diseases of the Liver, 1845.
[5] Edinburgh Med. and Surg. Journ. 1854.
[6] Clinical Researches on Diseases in India. 2 vols.

Morehead in which no intestinal ulceration existed. Frerichs, moreover, is of the same opinion, although by no means denying that, in certain cases, dysenteric as well as other forms of ulceration of the bowels may originate phlebitis of the coats of the portal veins, and so induce hepatic abscess.[1] The abscess of the liver which is found in intimate connection with Dysentery is the multiple abscess, small but numerous collections of pus. This form of purulent deposition Dr. Murchison has very distinctly shown to differ from the ordinary abscess of the liver which occurs in warm climates. In the latter case there is but one abscess which may attain a very large size, or in a few instances there may be two or three collections. Thus the *pyæmic* or multiple abscess, which is the common form of hepatic suppuration in this country, is to be distinguished from the *tropical* abscess of India and other hot climates; and while the latter may co-exist with Dysentery, such connection is wholly accidental. On the other hand, the multiple hepatic abscess, although by no means of frequent occurrence in India, is sometimes met with, but only, as Dr. Murchison has pointed out, in connection with Dysentery or some other source of purulent absorption. The only marked instance of hepatic abscess in connection with dysentery which has fallen under the writer's immediate observation was that of a soldier in a Highland regiment, who, while serving in India, became affected by the latter disease, which ultimately assumed a chronic and inveterate form. He was ordered home, and during his voyage to England the liver became much enlarged. Greatly emaciated and reduced in strength, and still suffering from frequent loose stools, he sank shortly after reaching this country. Examination of the body after death revealed the existence of a very large number of small abscesses scattered throughout the entire substance of the liver, the tissue of which was in different parts the seat of considerable induration.

ETIOLOGY.—Neither in its acute nor chronic form is Dysentery now a common disease of this country. The decline in the frequency of its occurrence has also been accompanied by a diminution in the severity of its attacks. From producing a very considerable annual mortality, as was the case in the seventeenth century, Dysentery now occupies a very low place among the causes of death. Essentially a disease of hot climates, its prevalence is, in these, observed to depend to a considerable extent on meteorological changes, while in temperate climates Dysentery is emphatically an autumnal malady. The continued exposure of the body to an elevated temperature predisposes to the occurrence of Dysentery; this it does, in all probability, by an injurious operation on the mucous membrane of the whole alimentary canal leading to its increased excitability, and by disordering the function of the liver: thus exposed, the sudden reduction of temperature, which so frequently takes place in the night season of our autumns, acts as a direct exciting cause of the disease. Thus, while heat predisposes to Dysentery, cold excites it. Unwholesome food has a potent action in the production of Dysentery. In this way unripe fruits, or even the ripe fruits when inordinately consumed, also vegetables, acid wines, and impure water, have particularly been supposed to act. There can indeed be no doubt that most of the slight, and some even

---

[1] " Eine causale Abhängigkeit der Hepatitis von Darmverschwärung ist also keineswegs festgestellt, wenn auch die Möglichkeit nicht geläugnet werden darf, dass ausnahmsweise unter begünstigenden Umständen dysenterische und andere Darmverschwürungen Phlebitis der Pfortaderwurzeln und hierdurch Leberabscesse erzeugen können."—*Klinik der Leberkrankheiten*, Zweiter Band, p. 113.

of the severer cases of Dysentery which we meet with, are occasioned by a distinct error in diet, or are traceable to the introduction into the alimentary canal of some substance or fluid of a deleterious or directly irritating nature. The not unfrequent connection of Dysentery with ague, and their observed alternation, have led to the impression that the former disease, like the latter, acknowledges an origin in malaria. That Dysentery may be produced by exhalations from putrid animal and decaying vegetable substances may perhaps be admitted; but the probability is that the relation of this disease to intermittent and remittent fevers, formerly insisted on, was not, strictly speaking, etiological, but to be accounted for by the disordered state of the portal circulation, which, occurring in ague, led indirectly to the inflammatory affection of the colon. The contagious nature of Dysentery has been asserted by some authorities; facts are, however, entirely wanting to prove the communication of the disease from person to person, in the sporadic form of the disease, with the consideration of which we are occupied; and in regard to the epidemic Dysentery, it may be admitted that the experience which appears at first sight to justify this conclusion, admits of another and more satisfactory explanation.

TREATMENT.—Dysentery in its acute form demands an energetic treatment; it is not a disease which can with safety be entrusted to the " vis naturæ medicatrix." Confinement to bed is of primary importance, the very rest favoring the arrestment of the malady, as much as movement of the body promotes its progress. Blood-letting was formerly practised in the treatment of Dysentery, and when pain is severe, and continues unrelieved by warm applications and rubefacients, local blood-letting by means of leeches applied over the track of the colon may still be had recourse to. The application of a few leeches to the verge of the anus has been recommended by some authors, and in the experience of the writer has appeared to be beneficial.

An indication of great importance in the treatment of Dysentery is to free the bowels from all irritating accumulations. This is best done by the employment of the gentler laxative medicines. Strong cathartics are not to be used. Castor-oil has been almost universally regarded as the best remedy for this purpose. Where much pain exists the oil may from the first be combined with a little laudanum; in the more advanced stages of the disease it will be prudent to associate the latter with it at every dose. The alternation of laxatives and opiates in the treatment of Dysentery has been highly praised by many practitioners. " It is the practice of some physicians," writes Sir Thomas Watson, " to prescribe laxatives and opium together; but in this complaint it is better to alternate them." [1] Opium by not a few has been regarded as the " summum remedium " in this disease. It was the favorite remedy of Sydenham in meeting the formidable Dysentery of his generation, and it is in allusion to its efficacy that the " prince of English practical physicians " rapturously exclaims— " And here I cannot but break out in praise of the great God, the giver of all good things, who hath granted to the human race, as a comfort in their afflictions, no medicine of the value of opium, either in regard to the number of diseases that it can control, or its efficiency in extirpating them. . . . . So necessary an instrument is opium in the hands of a skilful man, that medicine would be a cripple without it; and whoever understands it well, will do more with it alone

---

[1] Lectures on the Principles and Practice of Physic, vol. ii.

than he could well hope to do from any single medicine. To know
it only as a means of procuring sleep, or of allaying pain, or of checking a diarrhœa, is to know it only by halves. Like a Delphic sword it
can be used for many purposes besides. Of cordials it is the best that
has hitherto been discovered in nature. I had almost said it was the only
one." [1] Opium may be administered either in full or in small doses, and
each of these methods has its supporters. It may be given alone, or com
bined with ipecacuanha in the form of Dover's powder. Ipecacuanha itself is again largely employed, and more especially of late years in India.
We say *again* largely employed, for it is worthy of remark that ipecacuanha, originally known as a medicine about the middle of the seventeenth
century, was first used as a remedy in Dysentery. Brought to Europe
from Brazil by Piso, and some time afterwards made the subject of experiment in Paris by Adrien Helvetius, it was long known as the " radix
anti-dysenterica," [2] the "pulvis anti-dysentericus." Subsequently to its
original employment in France, in doses from one to three drachms, it was
used in this country and its colonies by Sir John Pringle and other physicians, in doses varying in amount, that ordinarily given being a scruple.
More recently the names of Mr. Mortimer, Mr. Twining, Mr. Docker, and
several other Indian surgeons, have been identified with the practice of
exhibiting ipecacuanha in Dysentery. The therapeutic action of the
remedy has been variously ascribed to its nauseant, its diaphoretic, and its
laxative or purgative effects. The latter was the view entertained by the
distinguished writer Sir John Pringle. Dr. Maclean thus expresses himself in regard to it: "It is probable that ipecacuanha owes much of its
usefulness in this disease to its action as an evacuant. It is a blood depu-
rant of an effective kind. It appears to increase the secretion of the whole
alimentary canal, as well as of the liver and pancreas: under its use tormina and tenesmus disappear, and feculent evacuations are more quickly
restored than by any other known remedy." [3] Dr. Morehead has always
used ipecacuanha in Dysentery from a consideration of its efficacy being
due to its laxative action. This physician counsels the exhibition of the
ipecacuanha according to the plan of the late Mr. Twining, [4] viz. "·from six
to three grains, combined with blue pill from five to two grains, and extract of gentian from four to two grains, every third, fourth, sixth, or
eighth hour, and to continue it steadily till amendment takes place. The
proportion of the dose and the frequency of its repetition must depend on
the acuteness of the symptoms. The duration of the treatment, and the
gradual diminution of the dose and of the frequency of its exhibition,
must be contingent on the rapidity and permanency of the amendment.
It must also be kept distinctly in view that, whilst the treatment by ipecacuanha is being pursued, it is often necessary—according as the state of
the pulse or the uneasiness of the abdomen on pressure may indicate the
necessity—to apply leeches; and also—according to the character and
scantiness of the evacuations, and the greater or less fulness of the abdomen—to give castor-oil occasionally in moderate doses." The reliance on
the therapeutic action of ipecacuanha is most conspicuously exhibited,
however, in the plan of its use suggested by Mr. Docker, and adopted by

---

[1] Medical Observations : Dysentery.
[2] For an interesting account of the early history of ipecacuanha, see " Traité Théra-
peutique et de Matière Médicale," par A. Trousseau et P. Pidoux, vol. i. p. 606.
[3] Reynolds' System of Medicine, vol. i. article Dysentery.
[4] Researches on Diseases in India, vol. i. p. 560.

Dr. Maclean,· and now generally followed in India. "The patient should be at once ordered to bed, and as quickly as possible brought under the influence of ipecacuanha in large doses. Some insist on the propriety of first giving a full dose of Battley's sedative, tincture of opium, or a few drops of chloroform, with the intention of making the stomach tolerant of the remedy, and restraining nausea and vomiting. I believe that the sedative in some cases is useful, and acts in the manner just described. On the other hand, I have often seen ipecacuanha do its work well, and with little disturbance of the stomach, without opium. Should it be determined to premise opium, thirty drops of the tincture should be given, and in half an hour followed by from twenty-five to thirty grains of ipecacuanha, which should be given in as small a quantity of fluid as possible; a little syrup of orange-peel covers the taste as well as anything else. As already advised, the patient should be kept perfectly still, and abstain from fluid for at least three hours. If thirsty, he may suck a little ice, or a teaspoonful of cold water at a time may be allowed. It is seldom that under this management nausea is excessive, and vomiting is rarely troublesome, seldom setting in for at least two hours after the medicine has been taken. The abdomen should be covered with a large sinapism, or a sheet of spongio-piline sprinkled with a little turpentine after being wrung out of hot water. In from eight to ten hours, according to the urgency of the symptoms and the effect produced by the first dose, ipecacuanha in a reduced dose should be repeated, with the same precautions as before. All who have had opportunities of trying this mode of treating Dysentery can bear testimony to the surprising effects that often follow the administration of one or two doses of ipecacuanha given in this manner. The tormina and tenesmus subside, the motions quickly become feculent, blood and slime disappear, and often, after profuse action of the skin, the patient falls into a tranquil sleep and awakens refreshed. The treatment may require to be continued for some days, the medicine being given in diminished doses, care being taken to allow a sufficient interval to admit of the patient taking some mild nourishment suited to the stage of the disease." If the writer be entitled to express an opinion regarding the use of a remedy which he has had but few opportunities of employing in the treatment of Dysentery, but has very frequently prescribed in cases of depraved action of the chylopoietic viscera, he feels inclined to ascribe the eminent therapeutic virtues of ipecacuanha to its direct action on the secerning function of the liver.

The employment of mercury in Dysentery is as warmly defended by some practitioners as it is condemned by others. In all stages and forms of the disease Dr. Maclean deprecates its use, while Dr. Wood asserts that no remedial influence is more effectual in Dysentery than that of mercury. Anything like the production of profuse salivation is certainly to be avoided; and although favorably influencing the progress of Dysentery in some cases, chiefly through its action on the liver, it will generally be admitted that in ipecacuanha, and in the employment of mild laxatives alternately with opiates, we possess more efficacious and certainly safer remedies.

It is in the more chronic form of Dysentery that such powerful astringents as acetate of lead, sulphate of copper, sulphate of zinc, the Indian Bael fruit, hæmatoxylon, and the sulphuric acid, are chiefly useful.

Among alterative remedies copaiba and turpentine, creasote and nux

---

vomica, have been commended. Quinine will favorably influence the progress of malarial Dysentery, when employed as an adjunct to other remedies; and iron, in the form of the pernitrate more especially, is called for when fluidity of blood as evidenced by hæmorrhages and cutaneous petechiæ exist; just as in scorbutic Dysentery, when chronic, milk and fresh fruits are indispensable articles of treatment.

Enemata of warm water cautiously introduced into the rectum are frequently grateful to the patient, and are useful in the early stages of Dysentery in bringing away hardened scybalous masses, the continuance of which in the bowels is attended by much irritation and suffering. Opiate enemata, and those containing ipecacuanha, and various astringents, may sometimes be employed with good effects. In Dysentery assuming a typhoid or adynamic type, it is necessary to support the patient's strength by the exhibition of stimulants; but these are, as a general rule, not well borne in this disease, and should always be administered with the greatest degree of caution.

The diet in Dysentery is of much importance. When the disease is comparatively slight and unattended by serious febrile symptoms, most farinaceous foods may be allowed. When, however, the severer form of the disease is in existence, bland drinks are alone admissible: milk with lime water, or Carrara water, may be regarded as the chief article of diet, and generally speaking is the one most relished by the patient.

Great attention should be paid to preserving the cleanliness of the patient, the dress, and bed-clothes, and in keeping the atmosphere of the sick-room as pure as possible, impregnated as it must from time to time become with the offensive odor of the discharges. The use of Condy's fluid, of weak chlorine vapor, or of carbolic acid for this purpose, is invaluable.

Sponging the surface of the body with tepid or warm water is desirable, and is usually found most grateful by the patient.

It may be added in connection with the treatment of chronic Dysentery, that change of air is frequently more efficacious than the use of drugs. Removal to the sea-coast, or a voyage, is specially to be recommended.

A flannel belt round the abdomen is an article of clothing which the convalescent from Dysentery, as well as all those who are prone to suffer from this disease, should adopt and constantly wear.

DUODENUM.

By S. O. Habershon, M.D.

The symptoms which have been regarded by some writers as proceeding from disease of the duodenum have by others been referred to states of the liver, of the stomach, or of the pancreas.

My own observations, and the facts which I adduce in the following remarks, show that there are symptoms of disease justly attributable to this portion of the alimentary canal; and that in some cases we may, with care, satisfactorily diagnose that the duodenum is diseased. The peculiarities of its position and structure deserve our careful attention. Extending from the pyloric extremity of the stomach to the jejunum, it is about twelve inches in length, and may be divided into three nearly equal portions; the first is the most movable, is almost surrounded by peritoneum, and is horizontal in its direction; it may be called the pyloric or stomachic portion of the duodenum, for it is associated with the stomach in its diseases. The second is vertical in direction, closely fixed near to the crura of the diaphragm, and to the vena cava; it receives the common bile and pancreatic ducts generally by a single opening, and is hepatic in its morbid relations. The pancreas is situated on the left side of the second portion; and the vena portæ, the hepatic artery, and the branches of the pancreatico-duodenal artery are also in relation with it. The third is horizontal in direction, and is simply intestinal in its function; the pancreas is situated above it; in front the superior mesenteric vessels enter the mesentery, and behind it are placed the aorta and the vena cava. The three portions of the duodenum are situated on *different planes*, the first portion being near to the anterior abdominal parietes, whilst the third part is immediately upon the spine; and this arrangement allows the contents of the canal mechanically to gravitate quickly into the jejunum, and assists also the discharge of bile from the ducts.

The muscular layers of the duodenum are double; a circular and a longitudinal coat, as in other portions of the small intestine. The mucous coat is covered with villi, which commence at the duodenum, and soon become exceedingly numerous; so also the valvulæ conniventes are gradually developed, till we find them as large as in the jejunum. The whole of the surface is studded over with Lieberkühn's follicles; not unfrequently, especially in young subjects, there are solitary glands, as in the jejunum and ileum. There are also the glands of Brunner, minute compound glands peculiar to the duodenum, and which are situated beneath the substance of the mucous membrane; these commence a few lines from the pylorus, and extend about as far as the common bile-duct; their function is not definitely known, but they are believed to resemble minute salivary or pancreatic glands. It sometimes happens that the soli-

tary glands are so distinct that they may very easily be mistaken for
Brunner's glands; the latter are, however, situated beneath the mucous
membrane, and microscopical examination at once manifests their difference.

There is still another point in connection with the duodenum that
deserves consideration, and which indicates its close connection with the
stomach and with the liver. The pneumogastric nerves, branches of
which supply the stomach, and also the liver, send filaments along the
first portion of the duodenum, continued onwards from the lesser curvature of the stomach; this associates that part of the duodenum very intimately with the stomach. Besides this nervous supply we have, according
to the observations of Meissner and Auerbach, minute plexuses of nerves
both in connection with the mucous and muscular coats.

The pancreatico-duodenal artery, which supplies the greater part of
the duodenum, is from the hepatic, and the pyloric branch of the coro-
nary extends into the first part of the duodenum, so that in the arterial
supply we find the same association.

State of secretion.—The secretion is stated to be alkaline, and such is
probably the case; the acid reaction after death arising from the gastric
juice, which has gravitated through the pylorus. Whether a patulous,
feeble contractile power in the pylorus, allowing the secretions of the
stomach to pass at irregular periods into the duodenum, is the cause of
the discomforts associated with these forms of dyspepsia, we have no data
on which to form an opinion. Corvisart states that the pancreatic fluid
discharged into the duodenum has the power of dissolving albuminous
substances; this opinion is, however, controverted by Dr. Brinton; the
former describes duodenal dyspepsia as arising from an abnormal condition of this secretion.

Congenital malformation.—The duodenum sometimes has a double
sigmoid curvature—a peculiar arrangement which I observed in a patient
who died from intestinal obstruction. The ascending colon was adherent
to the sigmoid flexure, and the cæcum, twisted upon itself, was situated
in the left hypochondriac region. The person had been born at the seventh month, and the cæcum was preternaturally free.

In a cyclopean monster, I found the viscera of a double fœtus in a
single peritoneal cavity; a double œsophagus was united in a single stomach, with a large convexity extending across the abdomen; and a single
duodenum, placed vertically, received the biliary pancreatic ducts on
either side.

Diverticula are exceedingly rare as compared with those which arise
from the lower part of the ileum; but small pouches are more frequently
present, and they consist generally of mucous membrane, thus constituting a sort of hernial protrusion. In the museum of Guy's is one of these,
situated near the opening of the duct into the duodenum.

Some believe that the duodenum becomes distended with flatus, or
with retained chyme, as the result of indigestion; and where there is
mechanical obstruction, which we shall afterwards describe, this may be
the case. It is possible also that an enormously distended transverse
colon may impede the free passage of the contents of the third portion,
but such is problematical. The distention which has been supposed to
arise from the duodenum will generally be found to be distention of
the stomach or the transverse colon; for the duodenum passes quickly
to a lower level, and I believe its contents at once gravitate into the
jejunum.

As to the strictly pathological states, we find congestion sometimes active, more frequently passive; ulceration, cancer, and lastly mechanical obstruction are also noticed.

To some it may appear altogether futile to speak of congestion or hyperæmia of the duodenum, but observation of the appearances after death convinces me that marked changes occur, and that in some instances a careful investigation might have pointed out their existence during life.

Great congestion of the duodenum is found in various diseases in which a similar condition extends to the whole tract of the alimentary canal, as in disease of the mitral valve, and in portal obstruction in hepatic disease; but there are other cases in which we find active congestion, especially in acute pneumonia. The latter state of acute hyperæmia is illustrated in the following case:

CASE LXXXV.—*Inflammation of the Bronchi, of the Bile-ducts, or Biliary Hepatitis, &c. Acute Congestion of the Duodenum.*—Thomas H——, æt. 42, was admitted into Guy's Hospital March, 1852; he had been ill for three weeks. He was a large, stout man, who for fourteen years had been in the police service; his habits of life had been very intemperate. Four years previously he had received a severe blow in his right side from a prize-fighter, and for some time he had been subject to vomiting in the morning, and the bowels had at times been much relaxed; before admission jaundice came on; he had had more anxiety of mind than usual, and gradually became languid and icteric. For four days his legs had swollen, afterwards his abdomen, and his strength became prostrated. The skin was of a dusky yellow color; the tongue was dry, brown, and furred; respiration 44; the pulse 100, soft and compressible; the abdomen was much distended with flatus, and fluctuation could also be felt; the liver extended several inches below the ribs, and there was tenderness on pressure in that part. In the chest there were general bronchial râles; he was delirious at night, and slept but little; the motions were light in color, the bowels relaxed, the urine contained lithates and the coloring matter of bile. Three days after admission he was more prostrate, and was delirious; the pulse was very compressible; he had pain in the right hypogastric region, and on the following day he died.

On *inspection* severe capillary bronchitis was found; the larger bronchi were also diseased; they were somewhat congested, and contained yellow-colored tenacious mucus. The heart was large, and had around it a considerable quantity of fat; the right ventricle was thin; the left ventricle had undergone partial fatty degeneration. The valves were healthy, with the exception of slight thickening of the mitral. *Abdomen.*—There were several pints of yellow serum in the peritoneum; the intestines were considerably distended with flatus, and the liver extended several inches below the ribs. The duodenum contained *bloody mucus, the lining membrane was very much congested, and in some parts ecchymosed.* The lower part of the small intestine contained clayey fæces. There was a considerable quantity of fat in the omentum, and in the abdominal parietes.

The liver weighed 7 lbs.; its surface was smooth, and of a deep greenish-yellow color, and some veins were seen upon it; the acini were whitish in color. The section of the liver appeared coarse along the smaller branches of the vena portæ; the capillary vessels in Glisson's capsule were much distended, and some of them were quite turgid with blood.

The smaller biliary vessels contained tenacious mucus, and their lining membrane was congested; this state of the bile-ducts contrasted remarkably with the pale color of the veins. The cells of the liver were gorged with fat, some of them were distended with oil-globules; other hepatic cells appeared ruptured, and granules with oil-globues were dispersed upon the field of the microscope. The deep green spots did not present any cells, but only granular matter.

The larger bile-ducts were free, but the opening into the duodenum was very much congested; the gall-bladder was empty; the kidneys were large and congested; the spleen was firm, and contained several fibrinous masses.

The health of this man was much impaired by his intemperate habits, and his liver had probably been diseased for a considerable period. The affection of the chest came on subsequent to his admission into the hospital, and consequently after the jaundice. There was evidently acute disease of the smaller biliary tubes, as indicated by the congestion of Glisson's capsule, by the congestion of the lining membrane of the biliary tubes, and the tenacious mucus they contained; the hepatic structure was stained with bile. The bronchitis which subsequently took place was, perhaps, the cause of the fatal termination, and tended, doubtless, to increase the congestion of the mucous membrane. The very congested state of the duodenum near the entrance of the bile-ducts indicated an extension of disease from the duodenum to the bile-ducts, or *vice versâ ;* it was much more *localized* than is observed in the secondary congestion of the mucous membrane in pulmonary obstruction. This did not appear to be an affection in which much benefit could be obtained from the administration of mercury, but rather from salines with sedatives.

After burns the mucous membrane of the duodenum has been found greatly congested, and in several cases recorded by Mr. Curling in the 'Medico-Chirurgical Transactions' this part of the intestine was ulcerated. This statement has not been confirmed by the observations of Dr. Wilks, recorded in the 'Guy's Reports' for 1856. I witnessed many of the cases to which he refers; and although in some the first part of the duodenum was hyperæmic, in none did I observe ulceration. A case of ulceration of the duodenum after a burn has, however, been placed in the Museum at Guy's, by Sir Wm. Gull. The child survived twenty-five days, but died comatose; a small cicatrizing ulcer was found in the first part of the duodenum.

Since the former edition of this work was written three cases of ulcer of the duodenum after burns have occurred at Guy's.

In one the patient was admitted for an extensive scald, and died thirteen days after admission. The duodenum contained two small ulcers, one the size of a pea, the other of a hemp-seed, and Brunner's glands were swollen. The ulcers appear to have had nothing to do with the man's death.

The second, a male child, æt. 4, died nineteen days after a severe burn of the lower extremities. He was doing well, and the burn was healing, when three days before death he began to pass blood into the bed. A large ulcer was found in the duodenum, and the pancreatico-duodenal artery was opened. The child had also two small ulcers on its tongue, extending through the mucous membrane.

The last case occurred in a girl, æt. 13, who died from tetanus about thirteen days after an extensive burn. The stomach was ecchymosed, and immediately beyond the pylorus was a small ulcer with thick raised edges. The thickening was considerable, so as to cause a suspicion that the ulcer antedated the burn. There was irregular injection around it. The pathology of such cases is still involved in much obscurity. Embolism and necrosis of tissue from blood extravasation after congestion have been suggested, as we have already mentioned, in stating the hypothetical explanations of acute perforating ulcer of the stomach.

Mr. Curling describes diarrhœa, and the discharge of blood, as having arisen from this condition of the duodenum, and sometimes severe hæmatemesis and prostration. In some instances death took place from peritonitis consequent on perforation. After such severe injury to the skin, it is not surprising to find great disturbance of the circulation or of the internal organs, and especially of the mucous membranes, which are known to sympathize so closely with the skin; in some of these cases stimulants appear to have been administered freely, and these have probably conduced to this diseased appearance of the duodenum.

Chronic congestion produces gray discoloration of the mucous membrane; and in the examination of the discolored part we find that the deep color is produced by the deposit of irregular grains of pigment, very thickly placed in the substance of the mucous membrane, near to its upper surface, and probably in the coats of the capillaries; the apparent explanation of this state being, that gastro-enteritis, or long-continued hyperæmia, has been followed by the deposition of hæmatine or pigment in the substance of the membrane.

In several cases of this gray discoloration the appearance, both in children and in adults, has been uniform. A child, æt. 9, a thin, poorly nourished, pale boy, who had been subject for some time to looseness of bowels, whilst running, hurt his thigh; he shortly afterwards complained of pain at that part; he was admitted into Guy's in a typhoid state, and died two days afterwards. There was suppuration in the brain, and gray discoloration of the mucous membrane of nearly the whole of the small and large intestines.

Chronic congestion is observed, as before stated, in connection with pulmonary and hepatic congestion, in fact, in any disease which leads to distention of the vena portæ; and we also find a less general condition of congestion of the first part of the duodenum in disease of the pylorus, whether it be simple fibroid degeneration and hypertrophy, or true cancerous disease. The mucous membrane becomes thickened, its vessels congested, and its glands enlarged; sometimes, indeed, so much so that the glands might easily be mistaken for minute cancerous tubercles. The continued irritation thus leads to hypertrophy of the glands of the mucous membrane, as we find in other similar structures.

The duodenum is sometimes found, after death, to be filled with blood, and a coagulum is occasionally moulded into its exact form. This is due to extravasation of blood from ulceration and perforation of an artery, in the duodenum or in the stomach.

As to the symptoms arising from the conditions just described, they appear to be so continually bound together with those indicative of simple disease of the contiguous viscera, that definiteness and certainty cannot be attained. The vomiting and pain connected with hepatic disease and gall-stone are possibly due partly to the condition of the duodenum. In the latter there is probably spasmodic contraction of the canal; but of this we do not speak with certainty. In the cases described by Mr. Curling, vomiting was a frequent symptom; and the bilious evacuation in violent vomiting indicates that the first and second portions of the duodenum have been involved.

Instances are not unfrequently met with in which, several hours after food, there is pain at the region of the duodenum, perhaps with violent vomiting, faintness, pallor of the countenance; and these symptoms have by some persons been referred to the duodenum, as a form of *duodenal dyspepsia or inflammation ;* by others to the pyloric valve; but occasionally jaundice follows, which appears to strengthen the former supposition. After intemperance, also, violent bilious vomiting, a furred state of the tongue, loss of appetite and loathing of food, diarrhœa, tenderness of the right hypochondriac region, are followed by jaundice; and we are prone to regard the duodenum as being in, at least, a state of great hyperæmia. Exposure to cold, with great mental anxiety, tends also to promote this state of duodenal disease; and the mischief appears to be propagated to the bile-ducts. Sir H. Marsh has drawn attention to the occurrence of jaundice with disease of the duodenum, in the ' Dublin Medical and Surgical Journal ;' see also Dr. Stokes, in the 'Encyclopædia of Practical Medicine.'

Congestion of the duodenum is best relieved by diminishing portal and hepatic engorgement, and by stimulating the abdominal excretory organs to increased action. These objects may be attained by giving saline and mercurial purgatives, by aperient enemata, and by the application of leeches to the anus or to the scrobiculus cordis. A free dose of calomel, blue pill, or gray powder, followed by a saline aperient draught, often acts very effectively as a purgative; but in many instances, especially where the morbid condition arises from chronic pulmonary disease or obstructive disease of the heart, small doses of mercurials may be very advantageously combined with squills and foxglove, so as thoroughly to act on the abdominal excretory glands; but to give mercury so as to produce salivation, or to prescribe it in every instance where bilious fluid is rejected, appears to be an unwise course. The most bland nourishment should be given, and abstinence from stimulants should be enjoined; ice and cold drinks often afford great relief when vomiting distresses the patient. In acute hyperæmic states, salines, as the solution of potash, the bicarbonates of potash or soda, the carbonates or the citrate of magnesia, may be given with diuretics in effervescence or otherwise, as the individual case may require. But in chronic hyperæmia, where there is profuse secretion of mucus, more advantage will be found from the dilute nitric or nitro-hydrochloric acids, with laxatives, as taraxacum, or with cinchona, and from the old compound gentian mixture of the London Pharmacopœia.

The most *acute* form of *inflammation* is sometimes observed after the administration of poisons. In a case of poisoning by sulphuric acid, where several square inches of the mucous membrane of the stomach had been destroyed, the duodenum was found intensely congested, and covered throughout by a thin, adherent, diphtheritic membrane. In this case the vomiting and dysphagia disappeared on the third day, and the patient, though extremely prostrate, did not appear to suffer much from pain. Arrowroot, lime-water, and milk, &c., were administered, and for a week it was thought that the patient might rally. (See " Diseases of Stomach.") In ordinary practice, however, we do not meet with this form of disease.

Ulceration of the duodenum varies both in degree and extent; sometimes it is merely superficial, and is associated with other diseases, as in a patient who died from albuminuria with pericarditis, in whom the duodenum presented superficial ulceration, the result of erythematous or acute inflammation; or there may be chronic ulcer, resembling that found in the stomach, and presenting many symptoms in common with that disease.

Some duodenal ulcers have raised and thickened edges, with depressed centres, being evidently of slow formation. They are mostly found in the first portion of the duodenum; and since this part of the intestine is almost surrounded by the peritoneum, we sometimes have fatal peritonitis, produced by perforation, as in the stomach, the muscular and peritoneal coats being also destroyed by the ulcer; or adhesion takes place with the adjoining structures, as the liver and pancreas, &c. ; and these oftentimes constitute the floor of the ulcer.

Several cases have come under my own notice the early symptoms of which were exceedingly slight, till sudden and fatal peritonitis had been set up by perforation. In some instances these ulcers have been associated with violent vomiting, the persistence and aggravation of which were attributed to this diseased condition; this occurred in a young woman, aged twenty-four, who was admitted into Guy's Hospital with very urgent vomiting; the pulse was small and frequent; she was pregnant, and died in a short time from peritonitis; a small ulcer was found in the duodenum.[1] The vomiting was probably referred to sympathetic irritation from the uterine state; and a favorable prognosis would in many such cases have been given till the symptoms of peritonitis came on.

The second portion of the duodenum is, however, also liable to ulceration, as in a case preserved in the museum of Guy's, where the coats of the whole of the vertical portion on the pancreatic side were destroyed, and the pancreas formed the base of a large chronic ulcer, in the centre of which was seen the opening of the biliary and pancreatic duct. There was a small ulcer in the third portion of the duodenum; and peritonitis had been set up; the pancreas was enlarged. The patient was forty-four years of age, and had empyema; he became exceedingly emaciated before death, and suffered from vomiting as well as from melæna.

Ulceration is sometimes followed by constriction; and adhesions also frequently form between the first part of the duodenum and the gall-bladder; in some, ulceration extends from the gall-bladder into the duodenum, thus allowing the passage of calculi; and the gall-bladder is, in other cases, entirely obliterated.

Pain several hours after food, a sallow complexion, furred tongue,. feebleness of circulation, mental depression, nausea, and irritable bowels, have been ascribed to ulceration of the duodenum, but the facts do not fully warrant this conclusion. In the several instances we have observed there were no such indications; in some, the ulceration was associated with disease of the gall-bladder; in others, with chronic disease of the liver; and the predisposing and exciting cause of the hepatic disturbance had probably induced the duodenal mischief.

Ulceration of the duodenum must be remembered both as a source of fatal perforation and of intestinal hæmorrhage, as well as of hæmatemesis.

The treatment of these cases is similar in all respects to that recommended for corresponding gastric disease.

---

[1] Dr. Hodgkin on ' The Pathology of Serous and Mucous Membranes.'

8

CASE LXXXVI.— *Ulceration of the Duodenum.   Perforation.*—
George E——, æt. 30, a man of light complexion, and of steady and temperate habits, was admitted into Guy's Hospital, October, 1851.   He was by trade a surgical instrument maker, and accustomed, when at work, to exercise pressure against the umbilicus.   Four months before admission he had slight expectoration of blood, but it was doubtful whether it proceeded from the lungs or stomach.   On October 20th, whilst apparently in good health, he suddenly experienced severe pain in the abdomen; to use his expression, he was " doubled up ; " he fell down in a fainting state, and was taken into a druggist's shop, where ammonia and some castor oil were administered.   The pain was situated on the right side.   On admission, he was in a state of collapse; the pain of which he complained passed in the course of the ureter.   On the following morning he was exceedingly depressed, the skin hot, the abdomen tender, and there were the symptoms of general peritonitis; vomiting of coffee-ground fluid came on, and pulsation was felt at the scrobiculus cordis, which suggested the idea of aneurism.   He survived fifty-six hours.   On examination, the peritoneum was found to be intensely inflamed; lymph was effused, and castor oil was found floating in the peritoneal cavity.   At the first part of the duodenum, about one inch from the pylorus, an ulcer was found of the size of a shilling; and at its base there was a circular opening, the third of an inch in diameter.   In the stomach several small aphthous ulcers were observed, and two small ones were covered with coagula.   The remaining parts of the small intestine were healthy; so also the cæcum, colon, kidneys, spleen, and liver.

In the chest there were old pleuritic adhesions on both sides, especially on the left, where there was also a small vomica, with indurated lung, and thickened tubes.

The patient was only thirty years of age; and, as he believed, in good health, though evidently of feeble constitutional power, as indicated by the condition of the lungs and the previous hæmoptysis; he was doubtless phthisical, but the disease of the duodenum resembled, in its insidious character, corresponding disease of the stomach, and gave no previous indication of its existence.

The treatment of the patient, before his admission, precluded all chance of recovery; but such, unfortunately, is too frequently the case.   Brandy and castor oil, probably both, found their way into the peritoneal sac; and the necessary removal of the man, at first into a druggist's shop, then to his own home, and afterwards a considerable distance to the hospital, tended to induce increased extravasation and peritonitis; the judicious administration of opium prolonged life many hours.

As to the cause, the stooping posture at his work probably assisted to produce the disease; but this is involved in much obscurity.

The position of the pain did not point out the seat of the perforation; but this is only what has frequently been observed in cases of gastric ulcer; the pain was principally in the right iliac fossa, and it was believed that the ileum, or appendix cæci, had given way.

Mr. Travers, in the 'Medico-Chirurgical Transactions,' mentions a case of perforation of the duodenum, about a finger's-breadth from the pylorus, in a gentleman, aged thirty-five, who was strumous, but considered to be in good general health.   There was a large irregular ulcer in the first part of the duodenum, with a small perforation, which had led to fatal peritonitis and death in thirteen hours; the perforation took place a short time after a meal, the period at which such accidents are generally found to occur.

CASE LXXXVII.—*Chronic Ulcer in the Duodenum. Carcinoma of the Liver. Jaundice. Granular Kidneys. Obliteration of the Bile-Duct.*—George C——, æt. 46, was admitted into Guy's Hospital December 14, 1853, and died January 4th. For a fortnight he had had jaundice, vomiting, and typhoid symptoms, and for three months, after exposure to cold, œdema of the lower extremities had been present. In the liver there were from six to ten carcinomatous tubercles; the bile-duct was obliterated near its opening into the duodenum, and throughout the liver the ducts were very much distended; the cells of the liver were normal. In the first portion of the duodenum there was a chronic ulcer, about an inch in diameter, with raised thickened edges, but not cancerous in its character; the rest of the intestine was healthy; the kidneys were large, and their surface irregular and granular.

The disease in the duodenum was not discovered till after death; the cancerous condition of the liver, inducing pressure on, and obliteration of the ducts; and the albuminuria appeared sufficient to explain all the symptoms. The ulcer in the duodenum, however, was in a chronic and passive condition, but nothing was ascertained as to its cause; we suppose that intemperance increased it. We rarely find such a complication of disease as cancer of the liver, acute disease of the kidney, and the condition of the duodenum just mentioned.

CASE LXXXVIII.—*Strumous Disease of the Abdomen. Perforating Ulcer of the Duodenum and Cæcum.*—Jane B——, æt. 18, was admitted into Guy's Hospital February 19, 1860, and died October 4th. At first the most prominent symptom was vomiting, which was supposed to be hysterical; but after a time the abdomen began to swell, diarrhœa came on, and emaciation, &c., increased, and these signs indicated the presence of organic disease. *On inspection*, the body was much emaciated; the legs were œdematous. The pleura was opaque, from the recent effusion of lymph, and the lungs were studded with tubercle. The peritoneum was acutely inflamed; the intestines were reddened, and there was lymph upon them; there were tubercular masses upon the peritoneum, covering the liver. On withdrawing the cæcum a small collection of offensive pus was found at its posterior part, and the abscess communicated with the cæcum by means of an opening about an inch above the ileo-colic valve. At the seat of perforation was a transverse ulcer, the edges of which were injected; the ulcer was one inch in length, and the opening one-third of an inch. A few other ulcers were observed in the colon, but none were found at the termination of the ileum. The mesenteric glands were enormously infiltrated with cheesy deposit; so also were the lumbar glands. Behind the first portion of the duodenum, and close to the pancreas, was a collection of offensive pus in front of the spine. This abscess communicated with the first portion of the duodenum by an opening about a quarter of an inch in diameter; the ulceration of the mucous membrane was more extensive than the external opening; and near to the perforation was a second smaller ulcer involving the mucous membrane. The first portion of the duodenum appeared to be contracted. The stomach was healthy, so also the kidneys. The spleen contained a softening strumous mass. The liver also was fatty.

Although the history of this case is imperfect, I have introduced it as an illustration not only of the obscurity of strumous disease in its earlier stage, but as an instance of irritation of the duodenum and colon, followed by ulceration and perforation, and producing peritonitis, at first of a local, but afterwards of a general character. The perforations in both situations were not directly into the serous cavity; the abscess connected with the duodenum was close to the pancreas upon the spine, and the one in the colon was placed behind the cæcum.

In an interesting case of hæmatemesis under my care in Guy's Hospital in 1875, the hæmorrhage which proved fatal was supposed to have come from the stomach, but on examination after death, it was found that a large ulcer in the duodenum had perforated the intestine, and led into a sloughing abscess in the portal fissure, with which the vena portæ communicated by an ulcerated opening partially filled by clot; the common bile-duct and hepatic duct were also divided; the hepatic artery was obliterated.[1]

It is probable that this perforation of the duodenum was from without, as was also the case in a patient under my care in 1866. A woman, aged 46, died a few weeks after admission, and a large abscess was found on the right side of the abdomen in the neighborhood of the ascending colon, along which it extended to the duodenum, where it opened by a rounded aperture an inch beyond the pylorus. The stomach contained a little altered blood. The patient had also cancerous disease of the gall-bladder, which, however, had no apparent connection with the peritoneal abscess.

CASE LXXXIX.—*Gall-stone. Ulceration of Gall-bladder and Duodenum. Large Gall-stone impacted in the Jejunum. Death from Hæmorrhage.*—A. B.——, æt. 56, had suffered from loss of appetite and mental depression for some time, due to family anxiety and trouble. He was a strong, muscular man, rather stout, and he had generally enjoyed good health. On November 29th, after a late dinner, severe pain came on in the region of the stomach, and for several hours was very intense; there was vomiting, and the pain extended to the back. On the following day the intense pain had subsided, but left soreness at the stomach, at the scrobiculus cordis and in the region of the gall-bladder. He had no appetite, and the tongue was furred; a purgative was given and saline medicine. On December 2d he had become jaundiced; the pulse was good, but the tongue was furred; there was no appetite for food, but much mental depression. The symptoms of jaundice gradually lessened. On December 15th the urine was still deep in color, but the motions were less pale. He lost the pain at the stomach, regained his appetite, the urine became normal in color, and he was able about Christmas to visit his friends; the skin, however, did not completely regain its color. On January 12th he returned to town, feeling tolerably well, but during the night nausea came on. On Saturday, 13th, sickness supervened, and he took blue pill with colocynth; the bowels acted a little. On the 14th the vomiting persisted, and saline effervescing medicines were prescribed; in the evening vomiting of blood occurred mixed with acid fluid. On Monday, January 15th, I saw him in consultation. The stomach was very irritable; everything was at once rejected; the pulse was quiet, 80; temperature normal; the abdomen was full, but there was no tenderness; he complained of soreness across the abdomen, just above the umbilical region, and

---

[1] See 'Path. Trans.,' vol. xxvii., 1876.

hardness could be felt at the scrobiculus cordis, which was thought to be the left lobe of the liver; there was no fixed pain, and no evidence of hernia. Bismuth medicine in effervescence was given, and a dose of calomel with colocynth. On the 16th he was rather easier, but there was no action from the bowels; the pain increased in the afternoon; the calomel and colocynth were repeated, and an injection used. On January 17th there was still no action of the bowels; a dose of castor oil was followed by violent vomiting of brown acid fluid; no flatus was passed; the pulse was 80, temp. 98°, the respiration easy; the abdomen was full and supple, and tympanitic, there was soreness in the epigastric region; no peristalsis could be seen. It seemed evident that there was obstruction in the bowels; purgatives were not repeated, but a grain of opium was given, and a turpentine enema was used. On January 18th.—The opium given night and morning had relieved the sickness; a full injection of oil and afterwards soap-and-water produced a discharge of hard scybala. Still there was no free action from the bowels; the pulse was 80, temperature still normal, the abdomen as before; the urine was normal in color, tolerably free in quantity, sp. gr. 1017, and it contained a trace of albumen. On the 19th he felt better in the morning, but as he could not pass urine freely a hip-bath was allowed. About 4 p.m. faintness came on, and he again vomited blood. The patient became restless. Still there was no action from the bowels; no flatus was passed, but the urinary bladder being distended a catheter was introduced, and about a pint of urine drawn off. Ice was applied externally, and some was swallowed, and astringents given. Nutrient injections were used repeatedly. At 10 p.m. he had rallied; about a pint of blood mixed with acid fluid had been rejected. On January 20th, about 5 a.m., more blood with clots were vomited, but he again rallied. On the 21st he had return of vomiting several times; in the evening he got out of bed, again vomited blood, faintness followed, and he died about 8 p.m.

Post-mortem examination by Dr. Goodhart twenty hours after death. —Abdominal wall thickly coated with fat. On opening the abdomen, the omentum and liver were found adherent to the abdominal wall in front at the upper part. The jejunum was much distended and dark in color; on tracing the small bowel from the cæcum upwards, the ileum was small and paler till its upper part was reached. Here it was blocked by a gall-stone of black color, somewhat irregular in shape, with a facet at either end of its long diameter, and measuring about 1¼ × 1½ inches. It moved about in the bowel under external manipulation with considerably freedom, though it would not pass far, and it quite filled the canal. Below, the bowel was empty or nearly so, and above, it was considerable dilated, and contained clayey and brownish pultaceous fæcal matter. The mucous membrane where the stone lodged was superficially ulcerated in some parts. About three inches higher up was a smaller gall-stone more like a fragment than a distinct calculus. It lay loose in the intestine with some fluid, brownish fæcal matter, and was easily crushed between the fingers. Nothing else abnormal was found till the duodenum was reached. On raising the right lobe of the liver the first part of the duodenum was seen to be pulled upwards and adherent to the fissure for the gall-bladder, and to hide the gall-bladder from view. The latter was further concealed by the omentum, also adherent to the liver. To the right of these structures was a little treacly blood, about a drachm, lying close to the duodenum underneath the liver, but free in the peritoneum. Its position there must have been of recent occurrence, as it was not shut in by adhesions, and yet no peritonitis was present. Dissecting out the gall-bladder and the vessels of the portal fissure, it was found that the fundus of the gall-bladder, the cavity of which was much contracted, opened by a large hole into a shreddy cavity which contained blood of treacly consistency; this cavity also opened by a large and irregular aperture into the duodenum, immediately beyond the pylorus at its anterior part. The vessels of the portal fissure ran to the left and in front of the cavity external to the gall-bladder, and they were not implicated, with the exception of the main branch of the hepatic duct to the right lobe of the liver. This was quite destroyed, and the truncated extremity opened into the abscess immediately behind its junction with the duct from the other side to form the main hepatic duct; the cystic duct was also destroyed. All the other vessels were normal. The cystic artery of the pancreatico-duodenal, the splenic and gastric arteries were all quite sound, and so also were all the branches of the portal vein in the neighborhood. The source of the hæmorrhage could, therefore, only be attributed to a venous oozing from the surface of the ulcer in the gall-bladder and the duodenum, and the sloughing cavity outside. The liver substance was unaffected by the ulcerative action, which was quite external to the capsule of the organ. The liver was small, but quite healthy, except a slight excess of fat. The kidneys were rather large and coarse; the right contained a cyst; the spleen was pale but healthy. The lungs were emphysematous. The muscular fibre of the heart was fatty.

From the observations I had made in November I felt convinced that the patient had gall-stone, and I supposed it had passed, although one was not detected. In the last attack the hæmorrhage was different from that which we generally observe in gastric ulcer; the blood was poured out more gradually. The clinical history was not that of gastric ulcer, neither was the hæmorrhage such as we have in engorgement of the portal circulation. From its gradual character, I thought it probable that it arose from the duodenum and was venous in character. It was evident, also, that there was mechanical obstruction of the intestine, for purgatives were instantly rejected, no true action from the bowels took place, and no flatus was passed. It occurred to me that possibly a gall-stone, impacted high up in the small intestine, was the cause of the obstruction, and this opinion was confirmed by the post-mortem examination, and also that the hæmorrhage arose from an ulcer in the duodenum.

No peristaltic movement, although several times looked for, could be detected, and yet the gall-stone was pushed down to the end of the jejunum. It is true that the abdomen was covered by a thick stratum of fat, which would render the observation of movement more difficult ; again, the intestine was filled with blood, and it is possible that the peristaltic movements were very feeble on account of the hæmorrhage. Another circumstance of great interest was the comparative absence of pain, although an enormous gall-stone, more than an inch in diameter, had ulcerated its way through the gall-bladder, then outside the bile-duct, into the duodenum; there was soreness, but no severe pain and no rigor. This comparative absence of pain I have previously noticed in a case where a large gall-stone had led to fatal obstruction by impaction immediately beyond the duodenum.

The following is a table of the cases in which we have found ulceration of the duodenum.

Sex.	Age.	Disease or injury.	Cause of death.	Remarks.
F.	13	Burn.	Tetanus.	Thirteen days after; stomach.
M.	4	Burn.	Hæmorrhage.	Ecchymosed ulcer on the tongue.
M.	..	Scald.	Exhaustion.	Brunne glands swollen.
F.	30	Primary disease.	Portal pyæmia.	
M.	39	Amyloid viscera.	Scrofulous pyelitis.	
M.	..	Diseased knee.	Hæmorrhage.	
M.	55	Hydrocephalus.	Convulsions.	Hypertrophy and dilatation; stomach.
F.	55	Renal disease.	Large white kidney.	
F.	12	Disease of hip.	Hæmorrhage from ulcer.	
M.	30	Primary disease.	Perforation. peritonitis.	
M.	46	Cancer of liver, &c.	Exhaustion from cancer, &c.	
F.	18	Tabes mesenterica.	....	Abscess behind cæcum, &c.
M.	56	Gall-stone.	Hæmorrhage; gall-stone impacted.	Ulcer due to the passage of a gall-stone.

Cancerous disease of the duodenum.—It is far more frequent to find the duodenum secondarily involved, than to be itself the primary seat of this fatal form of disease. In many cases the disease appears to have commenced in the pancreas or in the adjoining lymphatic glands, or in the liver; and although cancer of the stomach and of the pylorus is generally very defined and ceases abruptly at the commencement of the duodenum, such is not constantly the case, for the disease sometimes extends onward into the pyloric portion of the duodenum. Again, it is oftentimes very difficult to state precisely in which part the disease has commenced. As to the symptoms, the earlier ones are often very insidious; and are more likely to be mistaken for hepatic disease than the early symptoms of cancer of the stomach; still the first indications are those of dyspepsia and malaise, sallowness of complexion, mental depression, followed by nausea, vomiting, and sometimes pain, several hours after food has been taken. The patient emaciates, and a hardness or tumor is felt about the cartilage of the tenth rib; a very difficult question then arises, as to whether it is the pylorus that is affected, or the pancreas, or the lymphatic glands. Pulsation communicated to the growth may suggest the idea of aneurism. In aneurismal disease the vomiting is a less marked symptom, and the pulsation more uniform; the pain also is often very intense. In primary pancreatic disease the tumor is generally more central; the evacuations have been found sometimes to contain fat,[1] and until pressure take place on the duodenum, or the disease extend to the stomach, and to the lymphatic glands, the symptoms are less pronounced. Pyloric disease is indicated by more persistent vomiting than we find in simple duodenal disease. Occasionally local ulceration, with chronic thickening, takes place at the union of the transverse and ascending colon, or cancerous disease may be developed at this site, and subsequently perforate the duodenum. (See "Cancer of the Colon.") The formation of adhesions with the duodenum in these latter instances sometimes causes partial mechanical obstruction; vomiting is produced, and thus the diagnosis is

[1] The observations of Bernard tend to show that this symptom would be a constant one, if the duct were always obstructed.

rendered unusually difficult; such was the case in an instance which we shall presently give. In all these maladies there is emaciation, pallor, cachexia. Lastly, we must refer to numerous diseases of the omentum and of the liver as complicating the diagnosis. Here, however, the difficulty is less; for in the former the tumor is more central, there is greater mobility, and the gastric symptoms are less marked; in the latter, hepatic cancer, the tumor is more strictly in the hypochondrium, and the enlarged gland may be often felt with tubera projecting from its surface.

The termination of cancer of the duodenum is generally one of progressive emaciation and cachexia. If enlarged glands press upon the bile-ducts, jaundice will be added to the symptoms; if perforation or sloughing takes place, local abscess occasionally forms, which, by giving resonance on percussion, adds increased difficulty in forming a correct diagnosis.

The treatment of these cases generally consists in trying to relieve the distress and pain of the patient, and in sustaining his exhausted powers. Anodynes are required—opium, morphia, chloroform, or its preparations; and bland, but very nutrient diet, and especially of a fluid kind, should be given. Stimulants assist in keeping alive the flickering flame of life. When great sallowness of the complexion comes on, or jaundice, it is very unwise to give mercurials; they hasten degenerative changes, exhaust the patient, without any mitigation of his sufferings, and tend to hasten the fatal termination.

CASE XC.—*Cancer of the Duodenum.* (Reported by Mr. C. Longmore.)—James R——, æt. 40, was admitted under my care into Guy's Hospital, June 23, 1858, and died July 5th. He was by trade a coachbuilder, and he had resided at Newington; his habits of life had been temperate, and with the exception of a slight winter cough, he had enjoyed good health till Christmas of the preceding year. The first symptom of which he complained was a shooting pain in the back and stomach; the pain at last became very violent, especially at night after he had finished his work; there were also moving pains in both sides, especially on the right, and in the testicles; he had neither cough nor vomiting; about four weeks prior to his admission swelling of the feet came on, and after a few days his abdomen began to swell. He was a man of sallow complexion, with dark hair and eyes; he was much emaciated, but the feet and legs were anasarcous; there was dulness on percussion at the sides of the abdomen, and fluctuation was indistinctly felt. In the scrotum on the right side was a large hernial protrusion; and in the abdominal cavity a hard tumor could be felt, situated on the level of the umbilicus, and two inches to its left side; the tumor was an inch and a half to two inches in diameter, dull on percussion, but there was resonance around it; on pressure very slight pain was produced. Over the cartilage of the tenth rib there was also a minute pea-like tumor. The thoracic viscera were apparently healthy; the pulse feeble, compressible, 70. The surface of the body was cool. The tongue was coated with a brown fur in the centre, but was red at the tip. The bowels were freely acted upon, and the evacuations were paler than natural. The urine was scanty, sp. gr. 1032, free from albumen, but loaded with lithates. Small doses of acetate of morphia were given, and dilute nitric acid with infusion of cusparia. On June 25th the abdomen had greatly increased in size, it was very tense and resonant on percussion, except in the lumbar regions. On the 26th, the report states that, during the previous evening and on this day, he vomited about two quarts of bitter bilious fluid, but became more comfortable after its rejection; although a sensation of intense thirst came on. On the 28th he had become jaundiced, and complained of great pain across the loins, of an aching, dragging character.

On the evening of the 3d July vomiting of coffee-ground substance came on, and continued till his death on the 5th, at 11 p.m. The tumor several days previously seemed larger and more distinct. *Inspection* was made sixteen hours after death. There was rigor mortis; the whole body was jaundiced; the tissue of the heart was pale and softened. The liver was much enlarged. A tumor about the size of the fist surrounded the vessels at the fissure of the liver; the duodenum was situated in front of this growth, and was adherent to it. The commencement of the duodenum was quite destroyed by cancerous ulceration, and a large slough occupied the position of the first portion. The interior of the intestine communicated with the cancerous mass beneath it; the cancer tumor was altered in structure, and contained blood. The gall-bladder was distended to about twice its natural size, and contained a few gall-stones. The hepatic duct was slightly obstructed. The vena cava was in several places penetrated by the cancerous growth. The whole liver was filled with cancerous tubera, which were rounded, vascular, and softened. The disease appeared to run more especially in the course of the portal vessels, as if its entry into the liver had been by Glisson's capsule. The cancer growth consisted of large nucleated cells. The pancreas, supra-renal capsules, and kidneys, were healthy.

Instances of this kind are often very difficult of diagnosis, as to the precise seat of the disease; the glands close to the duodenum were probably first affected; but, although really behind the duodenum, the intestine did not cause resonance, probably on account of its becoming early implicated in the disease. The subsequent symptoms arose from pressure on the bile-ducts and the vena portæ, and from the degeneration of the cancerous growth. Mr. John Dix, of Hull, has recorded a very interesting case somewhat allied to this; and in which there was a tumor apparently connected with the liver, but resonant on percussion. "The tumor was hepatic and malignant. It was softening down—sloughing, in fact; and in this process it had involved and laid open the duodenum, to which it was attached; and whence air had escaped into a circumscribed cavity formed by the tumor behind, and the abdominal wall in front, to both of which the transverse colon was adherent below, forming the lower boundary" of the resonant space. The patient, "Mrs. M——, aged fifty-five, was pallid, feeble, and emaciated; she complained chiefly of pain in the right side of the abdomen, with vomiting and other symptoms referable to derangement of the hepatic and digestive functions. She had suffered, before that time, from jaundice and gall-stones." She died in about three months after the first medical examination; but the resonance in front of the tumor remained till death.

Primary cancer of the duodenum is of rare occurrence; a patient, under my care in Guy's in 1872, aged forty-five, suffered eighteen months before admission from violent vomiting and purging; for a week he was jaundiced, and he gradually sank; the stomach and pylorus were healthy, but the first portion of the duodenum was occupied by a large cancerous growth as large as a cricket-ball, soft, milky, vascular, and invading the liver by direct extension.

Instances also occur of primary disease of the pancreas extending to

the duodenum, and we have witnessed such cases in which the mucous
membrane of the duodenum had become infiltrated with medullary cancer.
Cancerous cachexia is then generally well marked, but till the pylorus or
duodenum become involved, vomiting is not generally a prominent symptom. We have also seen the duodenum perforated in cancerous disease
of the cæcum, which had extended upwards; and in another case, one of
villous cancer of the bile-ducts, a large cyst had formed in the right side
of the abdomen below the liver, and opened into the upper third of the
duodenum by four separate ulcers.

Mechanical obstruction.—Other parts of the intestine are much more
liable to obstruction of a mechanical character than the duodenum. In
the course of several years we have observed, or have found recorded, isolated cases of this kind of obstruction, arising from the following causes:—

1. Peritoneal adhesions.
2. Gall-stones of large size, which having ulcerated through the coats
of the gall-bladder, have become impacted in the duodenum, and have led
to fatal obstruction.
3. Enlarged glands, infiltrated by cancer, compressing the second or
third part of the duodenum.
4. Diseased pancreas.
5. Hydatid disease of the liver, opening into the duodenum.
6. Foreign bodies.

It is exceedingly common to find, after death, that adhesions have
taken place between the *first* portion of the duodenum and adjoining viscera, either the inferior surface of the liver and gall-bladder, or the transverse colon; and, in many instances, the impediment to the free passage of
the chyme is so slight that no symptoms point to any disturbed function.
In the following case adhesions with the colon were followed, however,
by great distention of the first part of the duodenum; but there was also
some ulceration of the same part of the intestine; there was chronic ulcer
of the colon, and chronic as well as acute peritonitis, with strumous and
glandular disease, so that there was considerable difficulty in unravelling
the symptoms, which resembled those of organic disease of the stomach.
Still we believe that the pain and the vomiting, several hours after food
had been taken, were the result of this duodenal obstruction.

CASE XCI.— *Chronic Peritonitis. Acute Peritonitis. Tubercular
Deposit on the Serous Membranes and in the Glands. Constriction of
the Duodenum, and great Dilatation of its first portion. Small Ulcer in
the Duodenum. Large Chronic Ulcer in the Colon.*—William C——, æt.
38, was admitted into Guy's Hospital under my care, April 15, 1861.
He was a married man, by trade a cooper, and he had resided at Dock-
head. About seven years previously he suffered from severe pain at the
epigastric region; and for several years since that time he had had pain
at the same part, but less acute in its character. He had never had any
hæmorrhage from the stomach, but he had complained of slight pain in
the dorsal region, between the sixth and eighth vertebræ. Some years
before he had had violent vomiting; but since that time vomiting had been
slight, the attacks coming on some time after food had been taken. He
had had slight pyrosis, and acid taste after vomiting. The pain at the
epigastric region was not constant, but it was worse after food, and was
especially aggravated by constipation.

On admission he was very much emaciated, with a sallow complexion,
and on the forehead there was a bronzed condition of his skin; the skin at

the elbows was also slightly discolored. There was moderate tenderness at the scrobiculus cordis; the abdomen was rounded and supple; no tumor could be felt; the bowels were rather confined; the pulse was very compressible; the tongue was red in patches. No disease could be detected in the lungs or heart. The patient stated that the bronzed color of the forehead had existed for three years, and had been produced by exposure to the sun; the lower part of the abdomen was also found to be slightly discolored.

On April 20th the bowels were freely moved, and he had severe pain at the scrobiculus cordis; the pain was neither relieved nor modified by any change of position.

He continued in the same prostrate condition, without pain or vomiting, till June 11th, when violent pain and symptoms of acute peritonitis came on, and he sank on the 13th.

14th.—*Inspection.*—The body was very much emaciated. *Chest.*—On the left side the pleura was firmly adherent, and on tearing it away, rounded, yellowish tubercles, two to three lines in diameter, were found thickly covering the costal surface. The left lung itself was very small; but there were no tubercles in it. The right pleura was free from adhesions or tubercles, and the lung was also quite healthy. The heart and pericardium were normal. There were several yellowish-white tubercular masses in the glands in the anterior mediastinum. On opening the abdomen, the intestines were seen to be distended; and the enlarged transverse colon, extending from one hypochondriac region to the other, prevented the stomach from being seen. There were numerous peritoneal adhesions, especially at the upper part of the abdomen, the transverse colon, stomach, and duodenum being united firmly to the under surface of the liver. The coils of the small intestine presented considerable injection at their lines of contact; but neither was lymph effused, nor had the serous membrane lost its shining color. Numerous tubercles were present on the serous membrane; some were exceedingly small, others three or four lines in diameter, and they were situated on the intestines or on the peritoneal surface of the liver. The mesenteric glands were extensively diseased; and all the glands situated in the neighborhood of the pancreas, and near the origin of the thoracic duct, were enlarged, although it could not be demonstrated that the duct was compressed. The glands contained much cheesy and cretaceous matter, and some more recent semi-transparent deposit. On removing the transverse colon, the stomach was found to be distended, and an elongated sac was produced, partially contracted, about three inches from the right extremity; this sac was at first supposed to be from hour-glass contraction of the stomach, but, on opening it, the first contraction was seen to be pylorus, and the second enlargement was an enormously distended first part of the duodenum. The stomach and duodenum contained grayish-green fluid and mucus. The mucous membrane of the stomach did not present any abrasion, thickening, nor ulceration, nor was the pylorus hypertrophied; there was a little arborescent injection. The sac formed by the first part of the duodenum was capable of holding eight to ten ounces of fluid, and was also injected. Immediately beyond the pylorus was a small ulcer about five lines by three in size, its edges rounded and without any injection; it did not extend into the muscular coat. Three inches from the pylorus the intestine was narrowed, and there was a constriction resembling a second pylorus; there was no thickening nor cicatrix, and it appeared probable that the peritoneal adhesions had looped up the intestine. On the gastric side of this constriction there was a small pouch, capable of admitting the tip of the finger. The rest of the duodenum, the jejunum, and the ileum, were healthy, with the exception of one or two small ulcers with tubercular deposit on their peritoneal surface. Peyer's glands were healthy. The cæcum and appendix also were normal. In the ascending colon the solitary glands were very distinct, and at the commencement of the transverse colon were the remains of an old ulcer; for two to three inches the mucous membrane was irregularly destroyed and puckered, and of a gray color. The rest of the intestine was normal. The supra-renal capsules, the kidneys, and the liver, were healthy; two or three strumous tubercles were, however, situated on the peritoneal surface of the liver.

In mechanical duodenal obstruction from the *second cause*, impaction of a gall-stone, the symptoms resemble those produced by internal strangulation of the intestine, or by hernia, but vomiting is set up at a very early period, and is of a severe character. The vomited matters, however, cannot have a stercoraceous odor nor appearance. The diagnosis is generally obscure and difficult; but where the symptoms of the passage of a gall-stone, namely, intense pain in the hypochondrium, vomiting, and subsequent jaundice, are followed also by the symptoms of insuperable obstruction, the nature of the malady is sufficiently clear; but in the ulceration of a large gall-stone through the coats of the gall-bladder into the duodenum, the indications of disease may be so slight as to be almost overlooked, and the subsequent obstruction cannot then be distinguished from strangulation taking place high up in the intestine. The impaction of the gall-stone is generally found to happen near the termination of the duodenum, or in the upper part of the jejunum.

In obstruction from diseased lymphatic glands in the neighborhood of the duodenum, the occlusion sometimes becomes suddenly complete, and the symptoms are those of internal strangulation; but more frequently the pressure is less, and the symptoms are those which we shall presently have to refer to in connection with disease of the pancreas; thus, in an instance of femoral hernia after the intestine had been returned, the symptoms continued, and the patient quickly died. The third portion of the duodenum was then found to have become firmly impacted between two enlarged glands.

CASE XCII.— *Obstruction from Biliary Calculus in the upper part of the Jejunum, thirty inches from the Pylorus.*—The calculus is in the museum of Guy's. The case was under the care of Ebenezer Pye Smith, Esq., and is recorded in the 'Pathological Transactions' of 1854. The patient was a stout woman, æt. 62. She had good health till three months before death, when she suffered slight pain in the right hypochondrium, which continued a fortnight, unaccompanied by sickness or prostration. She recovered, but continued her usual sedentary habits; five days before her death she began to feel sick, and vomited bile in large quantities; the urine was moderately secreted. The vomiting increased in violence, but with only very slight pain in the abdomen; on the fifth day she became comatose. A calculus composed of inspissated bile, and measuring four and a half inches in the circumference of its long by two and a half in the circumference of its short axis, was found impacted about thirty inches from the pylorus. There was much fibrous tissue on the under surface of the liver; and an ulcerated opening extended from the gall-bladder into the duodenum, below the bile-ducts.

The case just recorded of gall-stone with hæmorrhage and obstruction is of a somewhat similar kind. An interesting case is recorded by Dr. T. S. Gray in the 'Transactions of the Clinical Society for 1873,' in which a large gall-stone led to obstruction and stercoraceous vomiting, but was subsequently discharged, and the patient, a man aged 40, recovered.

There are in these cases three symptoms which especially deserve attention, as guiding us to a right diagnosis, when viewed in connection with the previous history. The absence of abdominal distention, the early period at which vomiting takes place, with the character of the ejected matters, and the diminution in the quantity of urine which is voided.

The absence of distention of the abdomen is an important sign of occluded intestine in the early part of its course. In obstruction of the large intestine, or even at the lower part of the small, the abdomen becomes enormously distended, and the peristaltic movements can often be observed in spare persons through the parietes; this is especially the case in disease of the sigmoid flexure of the colon. The stoutness of the patient sometimes renders this sign less observable; again, where this duodenal obstruction exists with hernia, the diagnosis must necessarily be most obscure. As to vomiting, it comes on very early, and the matters rejected are bilious. In strangulation of the ileum, and obstruction of the colon, unless irritating purgatives are given, this distressing symptom may be considerably postponed; and when it does take place and is continued, the matters are of a stercoraceous character. Still, in acute peritonitis, as from perforation, the sudden bilious vomiting may greatly mislead us. Again, very violent bilious vomiting sometimes takes place in disease of the stomach, and in cerebral disease; but the signs of obstruction are then wanting.

Gall-stone produces intense pain in the region of the gall-bladder, accompanied with vomiting and constipation; this severe character of pain we do not find in intestinal obstruction, but it must be acknowledged, that when slow ulcerative absorption has taken place between the walls of the gall-bladder and the duodenum, a calculus so extruded is followed by less severe suffering than in ordinary cases of biliary calculus.

A very interesting case, under the care of Dr. Lever, is mentioned by Dr. Barlow in the 'Guy's Reports' for 1844:—The patient, aged fifty-one, a year before her death had the symptoms of gall-stone, and the bowels afterwards became constipated; a short time before her death excessive pain, vomiting, and constipation came on, with scanty urine and collapsed abdomen. The gall-bladder and duodenum were firmly adherent; the two upper thirds of the duodenum were contracted, thickened, and would only admit a common quill; about the centre of the ileum was a biliary calculus of the size of a walnut, partially sacculated.

With regard to the quantity of urine excreted being a sign of the seat of obstruction, as mentioned in the paper by Dr. Barlow, just referred to, he argues that the quantity of urine must necessarily be small, from the diminished fluid brought within the range of the absorbing surface of the portal veins; and thus there must be diminished supply to the heart and kidneys; but there is often a large quantity of fluid ejected by vomiting which would proportionately lessen the renal secretion. If the obstruction be incomplete, or low down in the intestine, the kidneys pour out a larger quantity, and the vomiting is also less severe.

Dr. Barlow has, in the paper previously cited, dwelt upon the importance of bearing in mind, that in ischuria renalis, violent vomiting, con-stipation, and scanty urine are sometimes present.

In diseased pancreas the obstruction is less complete, but it acts by inducing firm adhesions about the first and second portions of the duod-enum; and pressure is also exerted by the increased size and hardness of the pancreas, and by infiltrated glands. The symptoms resemble those of obstructed pylorus, namely, vomiting several hours after food, grad-ually increasing emaciation, with constipation; and these symptoms are slowly developed during several months. A tumor can generally be felt near the region of the pylorus.

The following very interesting case was regarded as one of cancerous disease of the glands in the neighborhood of the pancreas, and secondary implication of the stomach; for the vomiting took place three or four hours after a meal, as in obstructive disease of the pylorus; and the gen-eral symptoms resembled those of organic gastric change.

CASE XCIII.—*Disease of the Pancreas. Suppuration and Gangrene. Pressure on the Duodenum.*—James P——, æt. 60, by occupation a pub-lican, and resident at Camberwell, was admitted under my care on July 4, 1861. He stated that he had always enjoyed good health till four months prior to admission, when he was suddenly seized with severe pain in the region of the stomach, and with vomiting. The vomiting returned at intervals of three or four days, and came on several hours after food. Four years previously he had begun to feel slight pain at the region of the stomach, which came on every three or four months, but was relieved by taking a little cayenne pepper with brandy. He had not received any blow, nor had he suffered from any hæmatemesis. The pain was situated at the epigastric and umbilical regions, and extended to the spine; it was of an acute kind, and had not the gnawing character of pain often de-scribed by patients affected with ulcer of the stomach.

On admission he was very much emaciated, with an anxious counte-nance, sallow complexion, and sunken eyes; his skin was hot and dry, and he complained greatly of thirst; the tongue was furred, the pulse frequent and sharp, the respiration normal; he had slight cough, but it did not dis-tress him; and there was no evidence of thoracic disease by percussion nor by auscultation. The abdomen was contracted moderately, except at the lower part of the epigastric and at the umbilical region, where there was a rounded tumor, evident on visual examination. The tumor was dull and tender on percussion; no fluctuation could be felt, and it had slight pulsation anteriorly from contact with the aorta, but no general aneurismal thrill. There was resonance between the tumor and the liver, as well as between the tumor and the spleen; in fact, both the hypochon-driac regions were more than usually resonant. Pressure on the tumor produced a feeling of nausea; the bowels were constipated; and the ap-petite was very poor. His weakness compelled him to remain quietly in bed. The urine was high colored and scanty, and was free from albumen. Fluid food was ordered, and soda-water with brandy, and chloric ether, ℥x., with nitrate of bismuth, gr. x., in mucilage mixture.

July 5th.—He was slightly relieved by the medicine, but the vomit-ing continued; the ejected matters consisted of deep-green fluid, con-taining a large quantity of mucus, of squamous epithelium, and some nucleated cells (from gastric glands). These attacks of vomiting dis-tressed him greatly; every kind of food was rejected at once, but the medicine and ice partially relieved his distress; his prostration, however, increased; hiccough distressed him; and he had an offensive taste in the mouth.

July 8th.—He was extremely restless and prostrate, and the vomited matters were of very deep-green color. At 9 p.m. he was suddenly taken worse, and continued in great pain during the night. At 7 a.m. next morning he expired.

Inspection seven hours after death.—The body was very much emaciated. The thoracic viscera were healthy, excepting old pleuritic adhesions. The peritoneum contained some dirty gray fluid, and had in some parts lost its shining smoothness; the intestines were slightly distended. The sac of the lesser omentum was distended by a large abscess, which had constituted the tumor felt during life. On tracing the duodenum upwards, at its centre was found an œdematous portion bulging out, and containing fluid resembling that in the peritoneum; but there was no perforation. By dividing the peritoneum between the stomach and the colon, an abscess was opened; it had dense fibrous walls, about two lines in thickness, in some parts irregularly sinuous, and having several bands on its walls, the remains of occluded vessels. Above and partly in front of the abscess was the stomach; below was the colon, and at its superior, right, and inferior parts was the duodenum greatly distended, and its coil enlarged. The abscess contained dirty offensive pus, and at its posterior part was a black slough about two and a half inches in length; some concrete yellow matter was also found on its walls. The abscess rested on the spine, the crura of the diaphragm, and on the superior mesenteric and splenic veins as they formed the vena portæ. It extended on the left to the spleen. The pancreas for two to three inches towards the splenic extremity was healthy, but the rest of the gland was in a sloughy state, and constituted the black mass found at the floor of the abscess. The pancreatic duct existed in the centre, and degenerating gland tissue was observed under the microscope. The gland and duct were separated from their duodenal attachment. The common bile-duct was healthy, and its opening into the duodenum was free; but the gall-bladder contained numerous gall-stones about the size of peas. The liver and spleen were healthy. The stomach was very much enlarged and distended; it contained tenacious green mucus, such as was vomited during life; its mucous membrane presented numerous points of arborescent injection, so also that of the duodenum; but no direct communication with the abscess could be found, nor any ulceration of the surface.

The origin of the disease in this remarkable case could not be ascertained, viz., whether a pancreatic calculus had set up the abscess, or whether inflammation had been produced in the cellular tissue about the gland. No direct blow had been received, and the disease slowly advanced. Acute peritonitis, from the transudation of offensive purulent serum into the general cavity of the peritoneum, was the cause of the fatal termination.

Dr. Bright believed that the fatty motions which he found in some of these cases were indicative of disease of the pancreas; but this symptom has not been constantly observed in pancreatic disease, possibly from the duct being only partially occluded.

The course taken by *hydatid disease* of the liver is uncertain; sometimes it is towards the surface, and a rounded tumor is then felt on the anterior abdominal parietes; or it extends through the diaphragm into the lungs. In a case under the care of Dr. Rees, in Guy's, the cyst opened into the duodenum. Hydatids were both vomited and passed by stool, and the former symptom was very severe. The patient was exceedingly ill, and a friction sound was audible over the seat of the tumor, evidently from local peritonitis; the patient steadily improved after the evacuation of the hydatids by vomiting; the tumor disappeared, and he left the hospital; but after a few weeks intense peritonitis came on, and he quickly died. The remains of hydatids were found in the liver; and the duodenum, colon, liver and kidney, were firmly united by adhesions. A large abscess existed between these structures, and had led to the fatal peritonitis. No communication existed between the liver and the colon; and although the duodenum at its second part was firmly adherent, no direct opening could be found.

The patient was twenty-nine years of age, and had resided at Twickenham; he was temperate in his habits; for nine years he had suffered from so-called " bilious attacks," and from vomiting, with slight sallowness of the skin; five years previously he had had severe jaundice, which continued for three weeks. Eight months before admission his appetite became ravenous, but he lost strength and became emaciated; for seven weeks he had been confined to his bed from severe pain about the umbilical region; jaundice came on, but disappeared, and was followed by very severe pain in the right hypochondriac region, extending to the loins, and a rounded growth presented itself below the ribs on the right side.

A remarkable instance of mechanical obstruction in the duodenum, from a foreign body, is recorded by Dr. Blakeley Brown, in the 'Pathological Transactions' of 1851 and 1852:—A delicate young woman, aged eighteen, became gradually emaciated, and at last died from peritonitis. The stomach, duodenum, and upper part of the jejunum, contained casts composed of agglutinated and interwoven masses of string and hair.

Gastric Solution of Duodenum.—The mucus of the duodenum is frequently found in an acid condition after death, which is probably due to some of the gastric juice slowly gravitating through the pylorus; but in some instances the pylorus is so patulous, that gastric juice readily passes, and exerts its solvent power after death in the same manner as in the stomach. Such a state was found in a child who died under my care in Guy's.

CASE XCIV.—*Perforation of Duodenum after Death from Solution by Gastric Juice.*—William B——, æt. 4, was admitted July 16, 1856, and died on the 23d. He was an anæmic child, with large head; on admission he was in a semi-comatose state, and the pupils were widely dilated; he had occasional vomiting, but no convulsions; six weeks previously he had had measles, and one week afterwards hydrocephalus gradually became developed; he was in an almost hopeless condition on admission.

Inspection was made fourteen hours after death. The arachnoid was covered with a slight layer of lymph, so as to give it a greasy appearance, and at the base of the brain there was considerable sub-arachnoid effusion. The ventricles contained two ounces of fluid, of sp. gr. 1001. There were miliary tubercles in the lungs and in the bronchial glands.

In the stomach there was considerable gastric solution, the mucous membrane being destroyed; but in the duodenum the intestine was quite divided, all the coats destroyed, and the end of the first portion terminated in an irregular ragged margin. The contents of the stomach were found in the peritoneal cavity. There were tubercles in the mesenteric glands, and an isolated one in the kidney.

DISEASES OF THE RECTUM AND ANUS.

By Thomas Blizard Curling, F.R.S.

An acquaintance with the numerous disorders of the lower bowel is absolutely necessary to qualify the medical practitioner to form a right diagnosis and judgment of the diseases of adjacent organs, as well as of the alimentary canal. Thus, complaints of the rectum are liable to be mistaken for affections of the uterus and even of the bladder; a discharge from a fistula in ano has been supposed to proceed from the vagina. Patients have been treated for obstinate diarrhœa, when the actual disease has been stricture in the lower bowel, or a lacerated perinæum and sphincter; and obstructions referred to the abdominal intestines have been discovered when too late to exist in their pelvic termination. The following is a table of the diseases of the rectum and anus; they can be treated of only very concisely in the space allotted to this subject:—Congenital Imperfections; Hæmorrhoids; Prolapsus of the Rectum; Irritable Ulcer; Irritable Sphincter; Nervous Affections of the Rectum; Villous Tumor of the Rectum; Polypus of the Rectum; Fistula; Chronic Ulceration; Stricture; Cancer; Atony; Anal Tumors and Excrescences; Prurigo Ani.

Congenital Imperfections of the Anus and Rectum.—These may be classed as follows:—1. Imperforate anus, without deficiency of the rectum. 2. Imperforate anus, the rectum being partially or wholly deficient. 3. Anus opening into a *cul-de-sac*, the rectum being partially or wholly deficient. 4. Imperforate anus in the male, the rectum being partially or wholly deficient, the bowel communicating with the urethra or neck of the bladder. 5. Imperforate anus in the female, the rectum being partially deficient, and communicating with the vagina or uterus. 6. Imperforate anus, the rectum being partially deficient, and opening externally in an abnormal situation by a narrow outlet. 7. Narrowness of the anus. A few other congenital deviations have been observed, but they are of very rare occurrence, the seven forms enumerated above being those most commonly met with.

The classification of these imperfections is founded on states which can generally be recognized during life. Unfortunately the condition of the terminal portion of the intestinal canal, and its relations to the parts around, cannot be predicated with any certainty. In cases of imperforate anus, or of anus opening into a *cul-de-sac*, the intestinal canal may terminate in a blind pouch at the brim of the pelvis, the rectum being wholly wanting; or an imperfect rectum may form a shut sac, descending to the floor of the pelvis, or as low as the neck of the bladder in the male, or the commencement of the vagina in the female. It is known that the anal portion of the bowel is developed distinctly from the upper portion, and that the two afterwards approximate and unite, the diaphragm or septum disappearing

9

by interstitial absorption.   A failure in this process is the cause of the second form of congenital imperfection.   The cases of imperforate anus in which the rectum communicates with the urethra or vagina depend on the original existence of a cloaca, the malformation being due to an incomplete separation during fœtal life.   These conditions are the result of an arrest of development at different stages.   The blind pouch in which the intestinal canal terminates is sometimes connected to the anal integument, or to the anal *cul-de-sac*, by a cord prolonged from the bowel above.   These cases are not, like the preceding, the result of a non-formation of the rectum, but are produced by an obliteration of the bowel, which was originally well formed; the obliteration being a pathological change due probably to ulceration and adhesion which had taken place during intra-uterine life.

These imperfections of the rectum can be remedied only by operative measures which vary according to the nature of the irregularity; and this treatment unfortunately often fails in obtaining a vent for the fæces, or in securing a permanent and sufficient passage.   In cases of failure in reaching the bowel at the natural site, life may still be preserved by making an artificial anus either in the left loin or in the left groin.   For several reasons the latter is the best situation for the operation in infants.[1]

HÆMORRHOIDS.—The hæmorrhoidal veins distributed in the sub-mucous tissue at the lower part of the rectum communicate in loops, and form a plexus which surrounds the bowel just within the internal sphincter.   The veins are best seen when somewhat congested, their deep purple hue being very apparent through the thin mucous membrane with which they are in close contact.   The plexus is then found to be about three-quarters of an inch in length, and composed of veins of various sizes, arranged for the most part lengthwise and in clusters, being especially collected in the longitudinal folds of the rectum.   The plexus does not extend lower than the external sphincter, but veins branching from it pass between the fibres of the internal sphincter, and descend along the outer edge of the former muscle close to the integuments surrounding the anus.

These veins are very liable to become dilated and varicose, giving rise to the disease termed *hæmorrhoids* or *piles*.   When the plexus beneath the mucous membrane is thus affected, the hæmorrhoids are said to be *internal*.   When the veins beneath the integuments outside the muscle are enlarged, the hæmorrhoids are called *external*.   Both external and internal piles very frequently co-exist.

We may distinguish two kinds of external piles.   1. A sanguineous tumor.   2. A cutaneous excrescence or growth.   The sanguineous tumor consists of a softish elevation of the skin near the margin of the anus of a rounded form, and of a livid or slightly blue tinge.   On cutting into it we find a dark-colored coagulum enclosed in a cyst.   This kind of pile is generally single, and seated at the side of the anus, but a second may form at a subsequent period.   The second form of external pile consists of flattened prolongations of skin.   They are generally the chronic results of the first form, a projecting fold left after absorption of the coagulum having undergone further growth.   The cutaneous excrescence contains no clot, and no enlarged or varicose veins; but clots and dilated veins may often be found at its base.   There is sometimes only a single broad flat excrescence at the side of the anus; but there are often two, one on each side, and occasionally more.   Similar excrescences occur as the result of

---

[1] See "Observations on the Rectum," by Mr. Curling. 3d edit. p. 221.

irritating discharges from the bowel, and are common in stricture and ulceration of the rectum.

The changes in structure consequent upon internal hæmorrhoids vary a good deal. In general the lower veins of the hæmorrhoidal plexus are dilated irregularly, or into pouches, which are filled with dark compact coagula. A bunch of varicose veins crowded in the lower ends of the longitudinal folds produce prominent projections of the mucous membrane, and deepen the pouches between the folds. Two or three of these prominences unite so as to form a transverse fold just within the sphincter. After a time the mucous membrane and sub-mucous tissue become greatly hypertrophied. Thus are developed elongated processes of a polypus form, and projecting transverse folds. The arteries, which are abundantly supplied to the lower part of the rectum, enlarge considerably, so that the mucous membrane involved is not only thickened, but extremely vascular. Such are the changes found in dissection, but the description conveys only a faint and incomplete impression of the condition of the parts observed during life.

Internal piles seldom attract attention until they have become developed so as to protrude at the anus in defecation. They then exhibit a remarkable diversity of appearance according to their number, size, and condition. The protrusion may consist of only one good-sized pile, found usually towards the perinæum or front of the anus. A single pile, consisting of a bright red projecting membrane connected with a loose fold of integument, and readily extruded, often forms in young persons, especially women. More commonly, there are three distinct prominent growths differing in size, one at each side of the anus, and a third in front; the latter, the perineal, being the largest. In old-standing cases they may be more numerous. The distinction between the piles is commonly well-marked, but not always; for the piles sometimes merge into each other, the protrusion forming a circular prominence. The aspect of extruded piles depends much upon their condition, whether congested, inflamed, or constricted by the sphincter. In a relaxed condition of the sphincter, they form softish tumors of a red granular appearance; but when protruded and congested, they constitute large tense tumid swellings of a deep red color and smooth surface, which readily bleed. When hæmorrhoids of large size are fully protruded, the integuments at the margin of the anus become everted, and form a broad band girting the base of the tumors externally.

External and internal piles often co-exist, the sphincter forming a narrow band separating the two. But the two forms may merge into each other, the difference being recognized by the character of the covering, mucous membrane or skin, the line of junction being visible on the surface of the tumors. Internal piles are confined to the lower border of the rectum. They never occur, as has been asserted, higher up the bowel, so that when they are entirely removed there is very little liability to a recurrence of the disease.

Hæmorrhoids is a disease of middle and advanced age. They rarely occur before puberty, and but few persons in after-life altogether escape them. All those circumstances which determine blood to the rectum, or which impede its return from the pelvis, tend to produce this disease. In many persons there is a natural predisposition to hæmorrhoids, and this may be hereditary. The complaint, indeed, often occurs in members of the same family who inherit the local weakness of their parents. But a predisposition is more frequently acquired by sedentary habits, indul-

gences at table, and excitement of the sexual organs, which explains the
well-known circumstance that hæmorrhoids are more prevalent in the
higher classes of society than amongst the laboring population.   The latter take plenty of exercise, live a good deal in the open air, and are little
liable to constipated bowels.   Hæmorrhoids, though common in both
sexes, occur more frequently in males than females.   Few women bear
children without becoming in some degree affected by them; but the urinary and genital disorders of the other sex, combined with freer habits of
living, are still more fertile sources of piles.

The symptoms produced by hæmorrhoids vary a good deal in different
subjects, and in different stages of the complaint.   External piles cause a
feeling of heat and tingling at the anus.   A costive motion is followed by
a burning sensation, and the excrescence becomes slightly swollen and tender on pressure, so as to render sitting uneasy.   This congested state of
the pile may subside, or it may lead to inflammation and considerable enlargement of the hæmorrhoid, which then forms an oval tumor, red, tense,
and extremely tender.   The irritation produced by costive evacuations, or
by friction in sitting and cleansing the part, sometimes gives rise to ulceration on the inner surface of the pile, causing a sore which extends a little within the circle of the sphincter.   This is liable to occur particularly
to those growths at the margin of the anus which hold a middle place between internal and external piles.   The pain in these cases is rather severe, being a burning sensation lasting for some time after defecation.

Internal piles, when slight, may exist for years, causing little inconvenience besides slight bleeding after a costive motion, and occasionally
a feeling of fulness, heat, and itching, just inside the anus.   When small
they protrude slightly with the mucous membrane in defecation, returning
afterwards within the sphincter.   When of larger size, they always protrude at stool, and require to be replaced, the patient usually pushing
them up with his fingers.   In a lax state of the sphincters, and a loose
and hypertrophied condition of the mucous membrane from which they
spring, piles come down, even when the patient stands or walks about.
When thus exposed to view they appear very prominent, of a rounded
form, and often of a deep purple or violet hue, have a soft feel, and are
evidently very vascular, bleeding readily when handled.   If free from congestion, they exhibit a florid red color, with a rough, granular surface.
In consequence of the friction and pressure to which they are exposed,
their mucous surface becomes abraded, and furnishes a mucous discharge
tinged with blood which soils the linen.   They are often so sore that the
patient is obliged to lie down, sitting causing great uneasiness.

Persons frequently suffer no inconvenience from piles until, irritated
by a costive motion, smart purgation, or the excitement of wine, they
become congested and inflamed, and cause spasm of the sphincter muscle.
Patients then have what is termed an "attack of piles"—that is to say,
they suddenly experience a sensation of heat, weight, and fulness just
within the rectum, followed by considerable pain at stool, and sometimes
irritation about the bladder.   Piles in this state are liable to be strangulated and constricted by the external sphincter, and hæmorrhoids of large
size have been known to slough off, the patients being relieved of the complaint by a sort of natural process, after much pain and suffering.   In
general the extremities only of one or two of the larger hæmorrhoids
perish, and the patient, though experiencing relief, is by no means cured
of the complaint.

One of the most common symptoms of internal hæmorrhoids, indeed, that from which the name of the complaint is derived, is hæmorrhage, which occurs when the bowels are evacuated. The bleeding varies greatly in amount. Sometimes the motions are merely tinged with a few drops of blood; in other instances the quantity lost is considerable, several ounces being voided at stool. The bleeding may be irregular, occurring only after costive motions, or in certain states of health; or it may take place daily, going on even within the bowel, and producing the usual symptoms of derangement from continued losses of blood. The complexion becomes blanched, and the lips appear waxy. The patient loses flesh and strength, has a quick and small pulse, suffers from throbbings in the temples, palpitations and difficulty of breathing on making any exertion, and at length the legs and feet become œdematous. The character of the bleeding also varies; it is sometimes venous, sometimes arterial. There are persons who are liable to discharges of blood from the hæmorrhoidal veins either at regular periods or when from good living or want of exercise the habit is fuller than usual. In these cases from three to six ounces of blood, or even more, pass away at stool, following the evacuation, and the blood which is voided is of a dark color and evidently venous. Such discharges must not be rashly interfered with. I had under my care, a gentleman, seventy years of age, who had been subject to hæmorrhoidal discharges for many years, usually in the spring and autumn. After lasting a week or ten days they generally ceased, but not always, and when faint and weak from their continuance, he was in the habit of arresting them with cold-water injections. The discharges at length ceased, but in six months afterwards his urine became albuminous, and a year later he died suddenly after an attack of epistaxis. Periodical losses of this character relieve congestion of the liver and kidneys, help to ward off attacks of gout, and prevent fits of apoplexy, so that in many persons they are rightly regarded as safety-valves. Persons who suffer from internal piles sometimes experience a pretty copious discharge of blood from the rectum. The bleeding shortly ceases, and all uneasy symptoms subside. This hæmorrhage is also venous. The escape of blood unloads the congested parts and the patient gets relieved. But the bleeding which most commonly occurs from internal piles is undoubtedly arterial, taking place from arteries enlarged by disease. The vessels on the spongy surface of the mucous membrane readily give way when blood is determined to the part in defecation or when abraded by the passage of hard fæces. An artery of some size, exposed by ulceration, continues to pour out blood, weakening the patient, and giving rise to the symptoms above described. Sometimes a small artery on the prominent part of a protruded pile may be observed pumping out blood. That hæmorrhage of this character is good for the health is quite a mistaken notion, and it is important that the practitioner should distinguish the bleeding taking place as a consequence of local disease from that which arises from a constitutional plethora or congestion of the intestinal organs.

When piles are small, and cause but little inconvenience, the treatment is very simple. In all cases attention should be paid to the habits of living. Persons with this complaint should take wine in great moderation, if at all, and they are in most instances benefited by abstaining entirely from stimulating drinks. Many individuals never suffer from piles, except after taking a glass of spirits and water, or a few glasses of wine. Such persons should become rigid water-drinkers. Active exercise in the open air should be taken daily, and the patient should avoid sitting too long at the desk, because it is by prolonged sedentary occupation and neglect of the rules of health that hæmorrhoid complaints are induced, which explains why literary persons so often suffer from them. Chairs with cane seats are to be recommended. The bowels must be carefully regulated, so as to avoid hard and costive motions, as well as frequent actions. Irritating the rectum by repeated purging is more hurtful even than constipation. On the other hand, when the liver is congested, or its secretions are sluggish, and when the bowels are costive, a mild cathartic, by clearing the intestines, especially the large, unloads the congested vessels and relieves the piles. Lenitive electuary, rendered more active when necessary by the addition of the tartrate of potash, will probably answer the purpose. The foreign mineral waters, the Püllna or the Friedrichshall, taken in the morning, fasting, agree well with many patients, and ensure a comfortable relief. When the intestines require fully unloading, a draught containing rhubarb powder and the tartrate or sulphate of potash answers without producing local irritation. Half a pint of cold spring water thrown into the rectum in the morning after breakfast has a very beneficial effect on the hæmorrhoids by constringing the vessels and softening the motions before the usual evacuation. The relief afforded by this treatment, combined with care in the mode of living, is often remarkable. Ordinary venous bleeding may be stopped in this way, using iced water, or some astringent such as a solution of tannic acid or infusion of rhatany. When the bleeding is of an arterial character, astringent injections are not so successful, and operative treatment often becomes necessary. When there is a slight slimy discharge from the surface of an exposed internal pile, benefit may be derived from the application of mild citrine ointment or the application of the solid sulphate of copper to the part.

External piles, when large and troublesome, and internal, when of such a size as to protrude at stool, and to be subject to inflammation, ulceration, and frequent bleeding, can be removed only by operation.

PROLAPSUS OF THE RECTUM.—In describing the changes occurring in piles, it was remarked that internal hæmorrhoids slip down and project at the anus. The descent of these growths is often attended with more or less eversion of the hypertrophied mucous membrane of the lower part of the rectum. In relaxed states also of the sphincter muscle and coats of the bowel, loose folds of mucous membrane are liable to protrude and to require replacement. This protrusion and exposure of the thickened mucous membrane with or without internal hæmorrhoids has been erroneously described by writers as prolapsus of the rectum. In the true prolapsus, however, there is a great deal more than an eversion of the internal surface of the bowel. The gut is inverted; there is a "falling down" and protrusion of the whole of the coats—a change in many respects analogous to intussusception, but differing from it in the circum-stances that the involved intestine, instead of being sheathed or invagin-ated, is uncovered and projects externally.

The length of bowel protruded in prolapsus varies greatly, from an inch to six inches or even more. The shape and appearance of the swelling depend partly upon its size, and partly upon the condition of the external sphincter. When not of any great length, the protrusion forms a rounded swelling which overlaps the anus, at which part it is contracted into a sort of neck. In its centre there is a circular opening, communicating with the intestinal canal. An inversion of greater extent usually forms an elongated pyriform tumor, the free extremity of which is often tilted forwards or to one side, and the intestinal aperture assumes the form of a fissure receding from the surface of the tumor, owing to the traction exerted upon it by the meso-rectum. In a relaxed state of the sphincter the surface of the protrusion has the usual florid appearance of the mucous membrane; but in other cases it is of a violet or livid color, and tumid from congestion, the return of blood being impeded by the contracted sphincter. The exposed mucous membrane is often thickened and granular, and sometimes ulcerated from friction against the thighs and clothes. A thin film of lymph may be occasionally observed coating its surface. On examining the section of a large prolapsed rectum from a child, I found the coats of the protruded bowel greatly enlarged; the areolar tissue was infiltrated with an albuminous deposit, the muscular tunic hypertrophied, and the mucous membrane much thickened and dense in structure, especially at the free extremity of the protrusion. These changes account for the difficulty in reducing the parts, and in retaining them afterwards, so often experienced in the treatment of this complaint in children, the bowel having become too large to be conveniently lodged in its natural position, and, like a foreign body, exciting the actions of expulsion. The atonic and relaxed state of the sphincter muscle in these cases is well shown by the facility with which one or two fingers can be passed through the anus even in children.

Prolapsus of the rectum is observed most frequently in children between the ages of two and four, but is liable to occur at a later period of life. In infancy it is produced by protracted diarrhœa; the frequent forcing of stool so weakening the coats and connections of the rectum, and relaxing the sphincter, as at length to lead to inversion of the bowel. The straining efforts to pass water consequent upon stone in the bladder often give rise to prolapsus in early life. In adults the descent results chiefly from a weakened condition of the sphincter and levator ani muscles, and a general relaxation of the tissues of the part. The rectum being imperfectly supported by the perinæum, the eversion at stool gradually extends until an actual inversion takes place, and this may increase until it forms a protrusion of considerable size. Prolapsus in adults is more common in women than in men. In the former it results in a great measure from weakness in the parts consequent upon child-bearing.

The annoyance and inconvenience occasioned by a prolapsus of the rectum vary considerably under different circumstances. Thus the bowel may descend only in a very slight degree at stool, and disappear by a natural effort afterwards, or it may come down only occasionally, admitting of being easily thrust back, and, when returned, will remain in its place until an attack of diarrhœa or the effort to pass a costive motion causes it to fall again. Prolapsus sometimes occurs after every motion, and even when the patient stands or moves about, forming a large red unsightly tumor exposed to friction, feeling sore, soiling the linen with a bloody discharge, and required to be pushed back frequently during the day. Or the gut may be constantly protruded, being fixed so as not to admit of replacement. There are cases on record in which a prolapsed bowel has become strangulated and inflamed, and has even mortified and sloughed off, similar to what sometimes happens to an invaginated intestine.

Young persons generally outgrow this complaint by the period of puberty; and common as is prolapsus in early life, it is rather rare in grown-up subjects. I have known, however, of persons, who have had this disease in childhood, and lost it, becoming affected with a return of it in after-life from the effects of a diarrhœa. In adults prolapsus is commonly attended with a slimy discharge of mucus tinged with blood, and, in some instances, with troublesome bleeding. The hæmorrhage does not occur from any particular spot, but as an exudation from the congested mucous surface when the bowel is protruded at stool. As the cause producing the bleeding is constantly recurring, there is sometimes considerable difficulty in arresting it, local applications having little effect so long as the bowel continues to descend.

In children, irritability of the bowels and diarrhœa must be checked and disordered secretions corrected by suitable remedies. Attention must be paid to diet, and when the powers are feeble benefit will be derived from quinine or steel. In slight cases it will be sufficient to direct the nurse, when the rectum comes down at stool, to place the child on its face across her lap, and to return the parts by taking a soft cambric handkerchief or sponge wetted in cold water, in both hands, and by gentle but steady compression to push the protruded gut back into the pelvis. The relaxed state of the membrane may be corrected by administering regularly every evening an astringent injection, such as the decoction of oak bark with alum, the infusion of rhatany, or the muriated tincture of iron diluted. The child should also be kept at rest in bed, and be made to relieve its bowels in the recumbent posture until the strong tendency to prolapsus has been corrected. The chief difficulty is to retain the parts after they have been reduced. A piece of sponge or cotton wool, moistened in an astringent lotion, may be lodged at the anus and secured there by approximating the buttocks by means of a broad strip of adhesive plaster applied across from one side to the other, and further secured with a T bandage. When the surface of the prolapsed bowel is ulcerated, it should be painted with a solution of nitrate of silver. In cases of stone, the prolapsus generally disappears after lithotomy.

Prolapsus in the adult requires surgical treatment to contract the opening of the anus by escharotics or operation. In old and unhealthy subjects the trouble may be remedied by a well-fitted rectum supporter.

IRRITABLE ULCER AND FISSURE.—The mucous membrane of the lower part of the rectum is arranged in longitudinal folds, which disappear in the expanded state of the bowel. These folds terminate below at the external sphincter. Just within this structure and between the folds, the mucous membrane is slightly dilated, variously in different subjects, but in many to such an extent as to form small sacs or pouches. Beside these folds, and in the spaces between them, there is a series of short projecting columnar processes, about three-eighths of an inch in length, separated by furrows or sinuses more or less deep, which are arranged around the lower part of the rectum. In the evacuation of the rectum, foreign bodies or little masses of hardened fæces are liable to be caught or detained in the pouches just described. It is in these little sinuses thus exposed to irritation, abrasion, and rent, that a superficial circumscribed ulcer is formed. On examining the ulcer without distending the rectum, the lateral edges only being presented to view, the breach of surface has the appearance of a *fissure*—the term commonly given, but improperly, to the sore, which though often originating in a rent is obviously more than a mere cleft or fissure in the mucous membrane of the bowel. Such an ulcer may occur in any part of the lower circumference of the rectum, but is usually found at the back part. It is quite superficial, and though sometimes circular is generally of an oval shape, its long axis being longitudinal and its lower extremity extending within the circle of the extended sphincter. On tactile examination the breach of surface and size of the sore can be readily distinguished. With the speculum, the ulcer being fully exposed is clearly seen not to be a mere fissure but a superficial sore. The surface is of a brighter red than the surrounding membrane, and has the usual indented appearance of an ulcer. A small pedunculated pile or polypoid growth, attached to the opposite side of the bowel, is frequently found in these cases. The growth lodges in the ulcer, adding to the irritation and the difficulty of cure.

The amount of suffering produced by this superficial ulcer varies a good deal, but the sore is generally extremely sensitive, and occasions severe distress. It is so situated that the fæces, in their passage outwards, rub over its surface, and the painful contact excites spasm of the sphincter muscle, causing a sharp burning pain, and often a forcing sensation, which lasts for two or three hours, the distress being usually greater after defecation than during the act; and in some instances, an interval, varying from five to ten minutes or more, elapses between the evacuation and the occurrence of pain. The pain is sometimes so acute that patients resist the desire to pass motions, and allow the bowels to become costive in dread of the sufferings brought on by evacuating them. I have known persons to deprive themselves of food in order to avoid an action. In one case, the intensity of suffering led the patient to adopt the dangerous course of inhaling chloroform whilst sitting on the close stool, and he could not be persuaded to go to the closet without this remedy.

The irritable ulcer occurs usually in middle life, and is more frequent in women than in men. It is met with as often in single as in married women. Though the symptoms are characteristic, the sore is often overlooked. On the attempt to separate the margins of the anus, or to dilate the sphincter to get a view of the ulcer or even to introduce the finger, spasm with an aggravation of pain is, in most cases, immediately excited, and the orifice becomes strongly contracted, and forcibly drawn in. When this is the case, it is better to desist, and to get an assistant to administer chloroform. Under its influence the sphincter yields completely, and the practitioner is able to ascertain the exact seat, character, and extent of the ulcer. In cases free from spasm, a good view may be obtained by simply dilating the anus with the two forefingers or by introducing a speculum.

The irritable ulcer seldom heals under the influence of local applications. The treatment necessary is an incision through its centre, including the superficial fibres of the sphincter muscle, in order to place this muscle at rest, to enlarge the passage and displace the sore; thus removing those sources of irritation which prevent its healing. An incision is not invariably required; but in all cases in which the pain is considerable, and in which there is much spasm of the sphincter, the attempt to procure the healing of the sore by local applications so often protracts the patient's sufferings, and so constantly ends in failure, that it is not desirable to make it. In cases complicated with a pedunculated pile or polypus, this growth must also be excised. When the suffering is moderate, a cure may be attempted by giving a laxative to ensure soft evacuations, rest in the recumbent posture, and the application of mercurial ointment with morphia, belladonna, or chloroform.

IRRITABLE SPHINCTER MUSCLE.—Persons occasionally suffer pain in defecation, especially during solid motions, increasing afterwards, and lasting half an hour or an hour. It is described as a forcing sensation, or a feeling as if the bowel were unrelieved. The anus is strongly contracted and drawn in by the action of the sphincter. Any attempt to examine the part induces spasm; and the finger passed through it is tightly grasped by

the muscle, as if girt by a cord. In cases of old standing, the muscle becomes hypertrophied and forms a mass, encircling the finger like a thick
unyielding ring. This irritability and hypertrophy of the sphincter sometimes produces serious trouble in defecation, owing to the expulsive powers of the bowel being insufficient to overcome the impediment caused by
this muscle to the passage of the fæces.

Irritability of the sphincter occurs commonly in hysterical females, or
in nervous susceptible women, who are accustomed to watch and to inten-
sify every sensation. The treatment required is mild laxatives, the local
application of an ointment containing chloroform, opium, or belladonna,
and the occasional passage of a bougie coated with a sedative ointment.
The bougie gives great relief in those cases in which an irritable sphincter
offers resistance to the passage of the fæces. In obstinate cases a partial
or complete division of the sphincter may be necessary to remove the
difficulty.

NERVOUS AFFECTIONS OF THE RECTUM.—The symptoms as well as the
causes of these complaints are usually obscure, and the diagnosis is often
perplexing. On analyzing the symptoms, they appear to consist, in some
instances, in an irritability, or a too frequent inclination to relieve the
bowels; in others, in a morbid sensibility or undue tenderness of the part;
and more rarely in an exaltation of sensibility independent of contact,
constituting neuralgia.

1. *Irritable Rectum.*—In derangements of the alimentary canal, and of
the organs connected with it, the fæces are often unhealthy and irritating
to the mucous membrane; consequently when passed into the rectum they
excite uneasiness, with an urgent desire to void them. Pressing and painful calls are also experienced when the bowel is ulcerated and in other
ways diseased. In "the irritable rectum " there is an inclination, more or
less urgent, to empty the bowel, usually at inconvenient times, although
the mucous membrane, as well as the fæces, are healthy, and often when
there is little or nothing to expel. Thus, a country rector experienced an
urgent desire to relieve the rectum in church, just before and during the
performance of divine service, notwithstanding an effort in the closet had
just previously proved ineffectual. He was subject to it also when attending public meetings and whilst riding in a railway carriage. Persons living in the country and going daily to business by railway are sometimes
annoyed by a desire to go to the closet just as the train is coming up, and
during the journey to town, but it passes off as soon as they arrive at the
counting-house and get engaged in business. The complaint is often connected with an anxious fidgety state of mind, against which patients may
often successfully struggle. My patient, the rector, got relief from a gentle aperient on the Saturday, and a mild opiate suppository administered
on Sunday morning.

2. *Morbid Sensibility of Rectum.*—Several cases have fallen under my
notice in which uneasiness has been experienced at a particular spot in the
rectum, being complained of, chiefly, during or after defecation. The fixity
and sometimes severity of the pain, and its aggravation from pressure, have
naturally led to the suspicion of the existence of some lesion in the mucous
membrane, such as an ulcer: but on careful examination, no breach of surface has been discovered; nothing has been observed except in some instances slight elevations and increased redness and vascularity at the spot
affected, and occasionally abrasion of the mucous membrane. The complaint consists chiefly in an exalted sensibility of the nerves of the part,
but the alterations in appearance just described indicate that there is also

some slight and superficial structural change. The remedies for the complaint are chiefly local. Sedatives, such as opium and belladonna, passed into the rectum give relief, but more permanent benefit may be derived from applications calculated to alter the character of the part, such as the sulphate of copper or a strong solution of the nitrate of silver applied through a speculum. I have in several instances cured severe morbid sensibility in this part by two or three caustic applications.

3. *Neuralgia of the Rectum.*—The two forms of nervous affection already described would be included by some writers under the general term of *neuralgia*, the sensibility of the rectum being in a measure perverted or augmented; but it will be remarked, that in the first no actual pain is experienced—there is merely an irregular and often causeless desire to evacuate the part; while in the second, the uneasiness consequent upon the augmented sensibility is either produced or aggravated by friction and pressure. In true neuralgia of the rectum, the pain is severe, but quite independent of contact. There is no tenderness. In the cases of neuralgia which have fallen under my notice, the pain was not characterized by paroxysms, by a suddenness of attack and disappearance, or by any regular intermittence, nor was the pain of an acute kind, but it was described as a continuous enduring pain, or a constant gnawing sensation, sufficiently severe to interfere seriously with the comforts and even the business of life. The pain was in no degree mental, for the patients were not persons of an anxious nervous temperament, and, unlike the two other forms of nervous affection, occupation and amusement had little influence in mitigating their troubles. The remedies calculated to give relief are such as are useful in neuralgia elsewhere, as quinine, steel, arsenic, bromide of potassium, local sedatives, and hypodermic injections, and they are as uncertain in removing the affection of the rectum as in curing neuralgia of other parts.

In some instances it is impossible to refer nervous complaints of the rectum to either of the forms just described, morbid sensibility and neuralgia being so combined as to prevent any distinction being drawn.

VILLOUS TUMOR OF THE RECTUM.—A growth similar to the villous tumor which occurs in the bladder and on other mucous surfaces sometimes forms in the rectum. It was first described by Mr. Quain under the name of a "peculiar bleeding tumor of the rectum;" but as it closely resembles the outgrowths found in the bladder called *villous*, I prefer the latter term. The tumor springs from the mucous membrane generally by a broad base, is soft in structure, and composed of a number of projecting papillæ or villi. On minute examination it is found to vary in structure according to the proportion of the fibrous or vascular elements entering into its composition. The villous tumor is innocent in character, and is not apt to return after complete removal. Its chief peculiarity in the rectum as in the bladder is a remarkable disposition to bleed. This growth is a rare disease, and occurs only in adults. When it projects at the anus, it exhibits characteristic projecting processes of a deep red color.

The bleeding to which this growth gives rise and the slimy discharge render its removal very necessary. If the tumor be attached high up, and a ligature can be applied round its base, this is desirable, as it would be difficult to arrest bleeding after excision.

POLYPUS OF THE RECTUM occurs in two forms—the *soft* or *follicular*, and the *hard* or *fibrous*. The soft polypus forms generally in early life. Its essential element is a considerable agglomeration of elongated follicles. There is a network of small vessels on its surface which is also furnished with papillæ. The polypus is attached to the mucous membrane of the rectum by a narrow peduncle which varies in length. The polypus is generally single, but several have sometimes been found. The follicular polypus usually makes its appearance external to the anus in children after a stool, and it resembles a small strawberry, being of a soft texture, granular on its surface, and of a red color. It produces no suffering, but causes usually a slight bloody discharge, which, occurring after every motion, excites attention. In some instances the bleeding is sufficient to weaken the patient. The description of the complaint by the mother or nurse is apt to mislead the practitioner and to induce him to conclude that the case is common prolapsus. The growth can generally be detected by the finger passed into the bowel; and when the peduncle is long enough, the tumor is forced out at stool, and its nature can then be ascertained without difficulty. The follicular polypus occurs very rarely in the adult.

The treatment of polypus in children is very simple and always effectual. The tumor should be strangulated by a ligature secured around the pedicle and then returned within the bowel. This causes no pain, and the polypus comes away with the motions two or three days afterwards. Excision is not quite safe, as it is liable to be followed by bleeding.

The fibrous polypus is of a pear shape, with a peduncle more or less long and thick. It varies in firmness, seldom bleeds, but occasions a slight mucous discharge; and when the peduncle is long, or the tumor low down, it protrudes at the anus after stool, and requires replacement. When lodged within the bowel, it causes a sensation of unrelief, as if a foreign body or feculent lump required discharge. The polypoid growth sometimes becomes congested, and when protruded in this state its peduncle is liable to become girt by the sphincter, which causes great pain. The suffering is still greater when, as frequently happens, the polypoid growth is complicated with an ulcer within the circle of the sphincter. The polypus, coming in contact with the ulcer, irritates it, and prevents its healing. The polypus must be removed by ligature or excision; and if an ulcer also exists, it must be divided at the same time.

FISTULA.—The loose areolar tissue around the lower part of the rectum is occasionally the seat of abscess, which bursts externally near the anus. But instead of the part healing afterwards like abscesses in other situations, the walls contract and become fistulous, and the patient is annoyed by a discharge from the opening. Such is the complaint termed *fistula in ano*. The abscess giving rise to fistula sometimes forms with all the characters and symptoms of acute phlegmon, suppuration taking place early, and the matter coming quickly to the surface. But more frequently a thickening appears at a spot near the anus with scarcely any sign of inflammation, and but little local pain, and is gradually resolved into a fluctuating swelling, which being opened discharges a fetid pus. On introducing a probe at the external orifice of a fistula formed in either way, it may pass through a small opening in the coats of the rectum into the bowel; the case is then called a *complete fistula*. When there is no internal opening, the complaint is named *blind external fistula*. The external orifice is usually but a short distance from the anus, its situation being often indicated by a button-like growth, and it is in the centre of this red projecting granulation that the opening is found. The aperture, however, is not always so marked, and being very small—a mere slit concealed in the folds of the anus—it cannot be detected without careful search. The abscess, before breaking or being opened, may have burrowed to some distance, and the external orifice may then be placed two or three inches from the anus in the direction of the buttock or perinæum. An abscess may make its way into the bowel before bursting externally, but the inner opening is generally formed after the external, and is small in size. The sinus burrows close to the mucous membrane of the rectum, which forms a thin barrier between the bowel and the sinus. Ulceration ensues at one point, and thus is formed the internal orifice of the fistula. The orifice is most commonly just within the sphincter: a fact established some years ago by M. Ribes, and fully confirmed by later observation. The inner opening, however, sometimes forms higher up the rectum, as I have clearly ascertained both in the living and dead subjects. Ulceration of the mucous membrane, from the wound of a fish bone or from other causes, may perforate the bowel just within the sphincter, and, allowing the escape of feculent matter into the areolar tissue around, may give rise to abscess and fistula. Fistula occurs in phthisical subjects, originating in tubercular ulceration of the mucous membrane and perforation of the bowel. In these cases the inner orifice is usually large in size, and there is sometimes a second opening. Though the inner orifice is most commonly found just within the sphincter, the fistula itself often extends some distance up the side of the rectum, as far as two or three inches, or even higher, and it may burrow in different directions. When the sinuses are tortuous, or pass in different directions, there may be more than one inner opening. Sometimes there is an external orifice on each side of the anus leading to fistulous passages which pass to the back of the rectum, and communicate with the gut at this part by a single orifice, so as to form a sort of *horse-shoe fistula*. The matter is liable to lodge in these complicated sinuses, to give rise to inflammation, and to lead to fresh abscesses and additional fistulous passages. In old-standing cases, the walls of the fistulous passages become dense and callous, feeling gristly to the finger. In all cases of complete fistula the occasional escape of a little feculent matter into the passage is amply sufficient to prevent the part healing, even if the actions of the levator and sphincter ani and the movements of defecation did not also interfere. Authors have described *blind internal fistula*, in which an opening into the bowel leads to a fistula without any external orifice. Such cases are rarely met with. The external opening sometimes closes for a short time, the spot being indicated by redness and induration; but sooner or later it re-opens, and the discharge returns, or a fresh opening is made at some distance off. It may happen, however, that the original ulcerated opening in the rectum being large, the matter from the abscess in the areolar tissue outside finds its way so readily into the bowel that the abscess does not burrow towards the surface. The situation of the suppurating cavity may be ascertained externally by a sort of hollow or indistinct fluctuating feel. A bistoury plunged into this will render the fistula complete. A blind internal fistula is very liable to be overlooked. I have met with several instances in which this has happened. In one case, the discharge, which was abundant and kept the linen constantly soiled, was supposed to proceed from the vagina.

An anal fistula is at all times an annoying complaint. Even when the seat of the disease is free from all inflammation and tenderness, the patient is troubled with a discharge which stains the linen and keeps the part uncomfortably moist. The discharge is usually a thin purulent fluid; at other times it is thick, and in complete fistula tinged brown from admixture of feculent matter. The discharge is more or less copious in different cases, and varies also at different times. It occasionally becomes so thin and scanty that the patient supposes the fistula is about to close, when he is disappointed by fresh irritation being set up, and the complaint becoming as annoying as ever.

Anal fistula is a disease of middle life, and occurs more frequently in men than in women. It is occasionally met with in young children, but rarely forms in advanced life, owing partly to the laxity of the rectum and sphincter in old people rendering the mucous membrane less liable to irritation and injury, and partly to the relief obtained by discharges from the hæmorrhoidal veins when congested.

The treatment necessary during the formation of the abscess, which precedes the establishment of a fistula, is rest in the recumbent posture, fomentations or the hip-bath, a poultice to the part, and mild laxatives. As soon as fluctuation can be felt, the prominent or central part should be punctured freely to prevent the matter burrowing in the loose areolar tissue, and thus to limit the extension of the sinuses. Fomentations and poultices must be continued until inflammation has subsided and the suppurating sac has become fistulous and indolent. An examination.may then be made. This, as well as the cure of anal fistula by operation, is entirely surgical.

CHRONIC ULCERATION OF THE RECTUM.—The rectum is subject to ulceration in dysentery and other diseases, the mucous membrane being destroyed to a greater or less extent. Chronic ulcers of a tubercular character also occur in this part, but they are generally small in size. Several cases of ulceration in the rectum, the origin of which must be ascribed to syphilis, have fallen under my notice, and this symptom is probably less rare than is commonly supposed. Syphilitic ulcers are usually large in size, and often involve the deeper structures of the coats of the rectum, so that the healing process is very apt to cause a serious contraction of the passage.

The chief symptoms referable to chronic ulceration of the rectum are— a purulent discharge from the anus more or less copious; motions generally loose and mixed or coated with a slimy fluid, and streaked with blood; soreness in passing stools and occasionally tenesmus. The pain in defecation varies considerably, being in some cases severe, in others very slight. Indeed, it is surprising how little suffering is often caused by the actions of the rectum and passage of the fæces in cases of large ulceration of the mucous surface. The suffering much depends on the position of the ulcer. Whether it be large or small, if it extends low down, so as to come within the grasp of the sphincter muscle, the pain is generally severe and persistent after defecation, and, in addition to other treatment, an incision through the lower margin of the ulcer is often required to release it from the actions of the sphincter.

The character, position, and extent of chronic ulceration in the rectum must be ascertained by examination with the finger and with the speculum. The surgeon will be able to feel a rough, uneven surface, more or less indented or depressed, and frequently hardness and consolidation of the walls of the rectum. The appearance of the sore in the lower part of the bowel may be seen through a glass speculum with an open end made oblique and large. This instrument is also very useful for the application of local remedies.

The treatment suitable to chronic ulceration greatly depends on the nature and extent of the disease, and upon the constitutional condition of the patient. In severe cases, I always keep the patient at rest in the recumbent position. In extensive destruction of the mucous surface with relaxed and copious discharges, especially when the disease originates in dysentery, vegetable astringents, such as simaruba, krameria, and bael, combined with the mineral acids and opium, are generally of great service in restraining the tenesmus and irritating evacuations and discharges. The subnitrate of bismuth with magnesia and anodynes often affords great relief. In many cases sulphate of copper with opium may be given with advantage. When the ulceration is consequent on syphilis or scrofula, the remedies appropriate to these diseases are required. The diet must be carefully regulated. The local treatment consists in the repeated application of weak solutions of nitrate of silver, and anodyne injections with mucilage, or anodyne suppositories.

STRICTURE OF THE RECTUM.—The rectum, like other mucous canals, as the œsophagus and urethra, is liable to obstruction from contraction of its walls, forming the disease called *stricture.* The contraction may be very limited in extent, and the stricture is then termed *annular;* or the contraction may include a portion, more or less considerable, of the bowel. The sub-mucous tissue is the chief seat of disease, and is condensed and converted into close-set fibrous tissue. The thickening of the coats of the bowel may be confined to part only of its circumference, or may be greater on one side than on the other, contracting the canal irregularly and forming a winding passage; or the induration, instead of being limited to a small portion of the bowel, may involve the greater part of the whole of the gut. The peritoneum investing the contracted bowel generally retains its healthy structure and appearance. Above the stricture the rectum is usually dilated and thickened. The enlargement results, not from a yielding of the intestine, but from a general hypertrophy of the walls of the bowel, and particularly of the muscular coat. The mucous membrane at this part is rarely healthy. It is red and tumid, or eroded and ulcerated, the diseased surface supplying during life a purulent discharge. There are often ulcerated apertures leading to fistulous passages which extend for some distance and open externally near the anus or in the buttock. The bowel below the stricture is generally more or less diseased, and frequently studded with small excrescences arising from partial hypertrophies or irregular growths of the surface and folds of the mucous membrane. These excrescences tend to narrow the canal below the stricture.

The seat of stricture in the rectum is at about an inch and a half to two inches from the anus, and easily within reach of the finger. In twenty-eight cases I found the stricture at this distance in twenty-one. In two in was nearer the anus, and in five at a greater distance. In three of the latter the stricture was at the point where the sigmoid flexure terminates in the rectum. In two instances I have met with double stricture.

The pathological changes causing stricture originate in chronic inflammation of the mucous and sub-mucous areolar tissue of the rectum. It is seldom possible to fix on the exciting cause, but it is well known that the part is exposed to numerous sources of irritation. Women, in whom the disease is much more common than in men, have sometimes ascribed its origin to a difficult labor, by which no doubt the bowel may be injured, so as to set up chronic disease. In twenty cases of women with stricture of the rectum I ascertained that the disease commenced shortly after a labor, and in some instances was attributed to an injury at that time. Injuries such as a kick, and violent use of an enema tube, have also been known to give rise to stricture. Strictures sometimes originate in the contraction consequent upon the healing of ulcers or wounds in the bowel, more commonly indeed than is generally supposed. In extensive dysenteric and syphilitic ulceration of the lower bowel the passage is liable to become seriously contracted in this way. I have met with several cases of stricture of this kind.[1] The rectum may also be obstructed by an outgrowth of fat, or by an infiltration of fat in the coats of the bowel. This is a very rare form of stricture. There is a specimen of it in the Museum of St. Thomas's Hospital, and Mr. Worthington has related a case in the *Transactions of the Pathological Society* (vol. xv.). In the Museum oi the London Hospital also there is a large fibrous and fatty tumor developed outside the rectum and contracting the passage.

Stricture of the rectum is a disease of middle life, being seldom met with in young persons except as a consequence of some injury. It is rare also in old people. The disease generally occurs between the ages of twenty and fifty.

The earliest symptom of stricture is, generally, habitual constipation with difficult defecation when the motions are solid. The difficulty being readily removed by a solvent purgative, the nature of the case is not usually suspected at this early period. As the contraction increases, the constipation is overcome with difficulty, and the patient acquires the habit of straining. The stools are observed to be small in calibre, and are often voided in small lumps. The mucous surface, irritated by the disturbance in the functions of the rectum, becomes inflamed and excoriated. This renders the action of the bowels painful, a burning sensation lasting for an hour or more after stool. There is also a secretion of brown slimy mucus, which escapes with the motions and soils the linen. The gases involved in the intestines not escaping readily, give rise to flatulent distention of the abdomen, especially in the course of the descending colon, and to disagreeable efforts for relief. The bowels often remain constipated for days together, and then a spontaneous mucous diarrhœa, excited by the fæcal collection or by a strong cathartic, softens the motions and enables the patient to void the accumulated mass, its passage being attended with pain. In other instances, the patient is teased with frequent fluid evacuations, and urgent desires to pass them. As the disease makes progress and ulceration ensues, the discharges become purulent and bloody, and the sufferings are much increased, the passage of motions being likened by the patient to a feeling as if boiling water was passing through the rectum. At this period, pain is often felt in the sacrum. The discharges are sometimes so copious that the stricture is overlooked, the case being mistaken for one of protracted diarrhœa. Ulceration often leads to abscesses and fistula, sinuses in the buttocks and labia being common complications of old-standing stricture of the rectum. The appetite and even the general health often remain good for a long time. The disease is very chronic; and so long as a passage for the motions can be obtained, the patient continues to follow his avocations, suffering more or less at different periods. The derangement of the digestive functions, the irritation kept up by the disease, and the exhausting discharges from the lower bowel in the course of time undermine the constitution and bring on hectic symptoms. The appetite fails, the body emaciates, profuse night-sweats ensue and the stricture directly or indirectly becomes the cause of death. This is sometimes hastened by a lodgment of hardened fæces, or of some foreign body just above the stric-

---

[1] See my " Observations on Diseases of the Rectum." Third edition. P. 119.

ture, so as to block up the bowel and occasion the symptoms of internal obstruction. Such an obstruction is sometimes the cause of an examination of the rectum, and thus leads to the detection of a close stricture previously unsuspected.

In order to detect a stricture it is necessary to make a tactile examination. On exposing the anus small flattened excrescences are usually observed at the margin of the aperture. These cutaneous growths resemble collapsed external piles, except that they are redder in color, and are kept moist by the escape of a thin discharge from the bowel. They originate in the irritation kept up by this discharge. The finger, well greased, being passed carefully and gently into the rectum, will be arrested on reaching the stricture, so that the point only can enter. If the contraction be somewhat recent and not very close, the finger may be carried with a gentle boring motion through the stricture so as to examine its whole extent. If the practitioner encounters much resistance or gives much pain, he must not venture to force the barrier, but must be content with ascertaining the seat and degree of contraction. In strictures high up in the gut, the rectum below may be found quite healthy, but it is often dilated and baggy with weakened expulsive powers. In strictures low down, the interior of the rectum is often abundantly studded with the small excrescences which I have described, which communicate to the finger the feeling of a number of rough irregular eminences, more or less hard, thickly lining the surface. This condition is invariably attended with a profuse discharge from the bowel of pus and slimy matter mixed with blood. A stricture high up in the rectum, and beyond the reach of the finger, is sometimes difficult of detection. In a suspected case the bowel must be explored by a flexible instrument. When the passage is free, a good-sized flexible gum elastic tube may always be passed into the colon. The point is apt to impinge on the sacrum, or to be caught in a fold of the bowel; but if some warm fluid, water or linseed-tea, be injected somewhat forcibly through the tube, a space is formed, admitting the easy transit of the instrument. In stricture, pain is felt when an instrument reaches the point of contraction, and a flexible one is arrested or passed on with more or less difficulty. In examinations for stricture it must be borne in mind that the rectum is liable to be compressed and obstructed by disease of the neighboring viscera—by an enlarged or retroflected uterus, fibrous tumors of this organ, a distended ovary, an excessively hypertrophied prostate,—an hydatid tumor between the bladder and rectum, or an outgrowth of fat, such as I have described.

The main object in the treatment of a stricture in the rectum is to remove the chronic induration and to dilate the contracted part sufficiently to admit a free passage for the fæces. The dilatation of the stricture is to be effected by mechanical means — by the passage of bougies, and sometimes by operation as well. The treatment, therefore, is chiefly surgical. An organic stricture fully established is universally admitted to be most difficult of remedy, and several high authorities, such as Dupuytren, Dr. Bushe, and Dr. Colles of Dublin, doubt the possibility of the disease being cured. These writers have undoubtedly taken too unfavorable a view of the results of treatment. In addition to the dilatation, means must be adopted to relieve the irritability of the part, to insure the regular passage of soft evacuations. An opiate suppository or injection may be lodged in the bowel at bed-time; and if the motions are costive, some confection of senna, castor-oil, or Püllna water may be taken

10

in the morning, in doses just sufficient to obtain an action of the bowels without purging. Castor-oil is often of great service. In small doses it softens the feculent masses, and lubricates the passage without weakening the patient. Cod-liver oil is also an excellent remedy. It nourishes the patient and softens the motions, rendering aperients unnecessary. The diet should be nutritious, and consist principally of animal food, so as to afford a small amount of excrementitious matter. It is no needless caution to advise patients to be careful to avoid swallowing plum-stones. Accumulations in the bowel above the stricture may be prevented by the occasional passage of an elastic tube through the contraction and an injection of soap and water. We sometimes meet, especially in hospital practice, with old, inveterate, and neglected strictures, in which the disease is too far advanced and the mischief too great to admit of relief by dilatation. In such cases, when the sufferings are severe, I have proposed the operation of lumbar-colotomy, and have performed it in two cases.[1]

CANCER OF THE RECTUM.—The coats of the rectum are subject to cancerous degeneration in the three forms of scirrhous, encephaloid, and colloid. The disease invades the coats to a greater or less extent, producing contraction of the canal, and it is liable to increase until it narrows the passage to such an extent that only a probe can pass through it. Fungoid growths sometimes spring from the mucous membrane at the side of the rectum and project into the bowel. Occasionally the bowel becomes blocked up and occluded by fungous masses. In other cases the changes which ensue have a contrary effect, degeneration and softening causing the coats to yield and increasing the calibre of the canal. A description of the progress of cancer of the rectum, and of the changes that occur in the advanced stage, is a description of the disorganization and invasion of all the tissues of the part, and of the organs in its immediate neighborhood, in various degrees in different cases. In some instances the carcinomatous bowel becomes wedged in the pelvis, agglutinated and fixed to the surrounding parts, forming one mass of disease. Frequently softening and ulceration cause fistulous communications with neighboring parts —with the vagina in the female, and with the bladder or urethra in the male; or the peritoneum may become perforated and an opening made into the abdominal cavity. When the passage is contracted, the intestine above becomes dilated and hypertrophied as in simple stricture. Carcinoma may attack any part of the bowel, but it generally affects the lower portion within three inches from the anus. It is liable to occur also, though less frequently, at the point where the sigmoid flexure terminates in the rectum. The disease is sometimes limited to the rectum and adjoining parts, though the lymphatic glands in the pelvis and lumbar region often become affected, the liver being invaded by tubercles and the peritoneum also studded with scirrhous deposits.

Cancer of the rectum generally commences insidiously. Its early symptoms are so similar to those of simple stricture, that the nature of the disease cannot be determined, or may not be suspected, until a considerable change has taken place in the condition of the bowel. The patient is troubled with flatulency, has difficulty in passing his motions, and strains in the effort to void them; and as the disease makes progress, he experiences pains about the sacrum, which gradually increase in severity and dart down the limbs. By this time probably some alarm is excited, and

[1] *Vide* London Hospital Reports, vol. iii.

an examination may be called for. The practitioner on introducing his finger into the rectum may easily detect a contraction more or less rigid; and should he feel any irregular nodules about the stricture, any hard solid tumor, or encounter a resistance like cartilage, or meet with softish tubercles which leave a bloody mark on the finger, then he would be able to decide on its being carcinomatous. At a later period no difficulty could be experienced. There is a hard mass of disease in which it may be difficult to discover the orifice of the passage, and sometimes round fungoid growths which bleed readily when touched. The disease may extend as low as the anus. An irregular red-looking growth sometimes protrudes externally, blocking up the passage or displacing the anus. The stools become relaxed and frequent and contain blood, and in passing cause a scalding pain and give rise to severe suffering. There is often a thin offensive discharge, and as the disease invades the sphincter, incontinency ensues. The loss of retentive power is often a great trouble in cancer of the rectum. This arises not only from the disease invading the anus and destroying the sphincter muscle, but occurs also when cancer is developed higher up in the bowel, the lower part being free. This may be explained by the carcinomatous disease pressing or destroying the nerves supplying the sphincter and so paralyzing it. The sufferings also increase. Severe shooting pains are referred to the groins, back, or upper part of the sacrum, and sometimes extend down the thighs and legs. The constitution suffers in due course. The patient acquires the blanched sallow look, anxious countenance, and emaciated appearance commonly observed in persons suffering from malignant disease. If complete obstruction does not accelerate a fatal termination, other troubles may arise. In consequence of a communication becoming established between the rectum and urethra or bladder in males, flatus and liquid fæces escape from the urinary passage, and in females motions are discharged from the vagina. The passage of part of the intestinal contents by these unnatural channels greatly increases the misery of the patient's condition, rendering him an object of disgust to himself and offensive to those about him. An ulcerated opening into the peritoneum, allowing the escape of feculent matter into the abdomen, may excite peritonitis and thus bring the case to a fatal termination; or the powers of life gradually giving way, the patient becomes hectic and exhausted, worn out by this painful and distressing malady. There is great variety, however, in the degree of suffering, and even of constitutional derangement, attending the disease. Whilst in some cases the sufferings are excruciating, in others they are comparatively slight. In my experience patients suffer less from the disease when developed high up in the rectum than when formed near the anus.

Cancer of the rectum occurs generally in middle life. The earliest age at which I have met with it is twenty, the patient being a young man in the London Hospital. It is commonly believed that this disease attacks women more frequently than men. This does not accord with my experience of cases seen in hospital and private practice. Of seventy-three cases of which I have preserved notes, fifty-seven were males and sixteen females.

All that can be obtained from remedies is palliation of the symptoms, ease from pain, and support under the wearing effects of this terrible disease. The patient should remain at rest, chiefly in the recumbent posture, and take a nourishing but not stimulating diet. The general health may be supported by tonics. The bowels must be kept open and the motions rendered soft by Püllna water or small doses of castor-oil. If the stricture

148   DISEASES OF THE INTESTINES AND PERITONEUM.

be close, injections may be necessary through a long tube to break up the feculent masses. The greatest care is necessary in the passage of the tube, as if force be used the carcinomatous mass may yield and the tube be driven into the abdomen. Bleeding may be checked by injections of sulphate of copper and tannic acid. Pain can be alleviated by opiate and belladonna injections, or by small doses of morphia taken night and morning, their strength being gradually increased as the effects of the remedy diminish. Subcutaneous injections of morphia also are effectual in giving relief. So great were the sufferings in a recent case, that after a time as much as 3½ grs. were thus injected twice a day.

In cancerous disease of the rectum attended with great suffering from incontinency and constant scalding discharges, I have advocated and performed in several cases colotomy in the left loin. By diverting the passage of the fæces, the local distress can be in a great measure prevented, and I have reason to believe that the progress of the disease also may be retarded by the removal of a source of almost continual irritation. I have established an anus in the left loin in several cases of cancer in which no obstruction existed, in order to mitigate the symptoms, with a satisfactory result in prolonging life and preventing suffering.[1]

EPITHELIAL CANCER OF THE ANUS AND RECTUM.—The anus, like other parts, where a junction takes place between the skin and mucous membrane, is liable to epithelioma. The affection is comparatively rare, and has seldom been noticed by writers. It is easily recognized by the ordinary characters of the sore. In the few cases which have fallen under my notice, the disease extended into the rectum, but there was no reason to doubt that its original seat was the anus. The only treatment applicable to this affection is caustics or excision. I prefer the latter, as more sure and thorough. Though more common at the anus, epithelioma may occur in any part of the mucous membrane of the rectum. When occurring up the bowel, the disease is apt to produce slight bleeding, but it is much less serious than scirrhous and medullary cancer. The latter produce sooner or later some contraction or obstruction in the passage, and show a tendency to involve the parts around. In epithelial cancer I have never noticed any impediment in defecation, and have invariably found the passage free and unobstructed. Neither do patients complain of the distressing pain, referred usually to the sacrum, which persons affected with scirrhus of the rectum so commonly experience, nor suffer painful tenesmus and defecation, which add so much to their distress in this form of the disease. There is also an absence of the cancerous cachexia, of the emaciation and pale and anxious countenance so frequently remarked in malignant disease. Epithelial cancer in the rectum may go on for years, but the patient becomes exhausted at last from repeated small bleedings. The hæmorrhage is best restrained by injections of solutions of sulphate of copper, chloride of zinc or tannin.

ATONY OF THE RECTUM.—In paraplegia the forces which expel the fæces and the retentive functions of the sphincter are both destroyed; consequently, the motions, if sufficiently liquid, on reaching the lower bowel escape involuntarily. I have not met with any well-marked case of paralysis of the rectum independently of palsy of the lower half of the body; but several instances of loss of tonicity or defective muscular power in the lower bowel, rendering it incapable of properly extruding its contents, have come under my notice. An atonic condition of the rectum

---

[1] *Vide* London Hospital Reports, vols. ii. and iv.

may be produced by the too free and frequent use of enemata, the quantity thrown up being so large as to dilate the bowel and impair the power of its muscular coat. This condition is apt to give rise to fæcal accumulations. Cases of this kind are not very uncommon, yet they are liable to be overlooked by practitioners. It appears that the rectum becomes gradually dilated and blocked up by a collection of hard dry fæces which the patient has not the power to expel. Some indurated lumps from the sacs of the colon, on reaching the rectum, perhaps coalesce so as to form a large mass; or a quantity accumulated in the colon on descending into the lower bowel becomes impacted there. In several instances a plum-stone has been found in the centre of the mass. Such a collection gives rise to considerable distress and alarm, producing constipation, a sensation of weight and fulness in the rectum, tenesmus and forcing pains. In cases of some duration, when the hardened fæces do not quite obstruct the passage, they excite irritation and a mucous discharge which, mixing with recent feculent matter passing over the lump, causes the case to be mistaken for diarrhœa. Injections have no effect in softening the indurated mass. They act only on the surface and return immediately, there being no room for their lodgement in the bowel. On digital examination the bowel is found to be distended and blocked up with a large lump which feels almost as hard as a stone. In such cases the only mode of giving relief is by surgical interference. The mass requires to be broken up and scooped out. Sir James Simpson has described this affection under the head of "ball-valve obstruction of the rectum by scybalous masses."[1] Some years ago I saw a lady who for eighteen months had been unable to relieve her bowels without aperients and without passing her finger into the rectum. On examination I detected a hard elongated mass which was forced down in the effort of defecation and obstructed the anus until the finger pushed it back. I broke up this mass, and after the bowels had been relieved by injections the difficulty was entirely removed.

ANAL TUMORS AND EXCRESCENCES.—Besides the flaps and folds of integument consequent on external piles, other growths are developed in the immediate vicinity of the anus. These tumors of a fibrous texture sometimes form in the subcutaneous areolar tissue, and as they increase become pedunculated. They seldom exceed the size of a chestnut, though I have known one to weigh half a pound. They have a firm feel, and their surface is in general irregularly lobulated. These growths may be easily and safely removed by excision.

Warts are not unfrequently developed around the anus, and they sometimes grow so abundantly as to constitute a considerable cauliflower-looking excrescence. They then form projecting processes of various sizes densely grouped together, many being of large size, with their summits isolated, expanded, and elevated on narrow peduncles more or less flattened. I have seen a mass forming a tumor as large as the closed fist, separating the nates, and almost blocking up the passage for the fæces. When abundant, they give rise to a thin offensive discharge. They originate in the irritation consequent on want of cleanliness, and occur generally in young adults of both sexes. I once saw a large crop of these growths in a child only four years of age. In some persons there is so strong a disposition to the formation of warts, that without great attention it is difficult to prevent their formation. If few in number and small in size, they may be destroyed with strong nitric acid. They usually re-

----

[1] Edinburgh Monthly Journal of Medical Science, April, 1849.

quire however to be removed by excision, which is the quickest and most effectual mode of treatment. Great cleanliness and the application of astringent lotions will be necessary to prevent their reproduction afterwards.

PRURIGO ANI.—Itching at the anus is a common symptom in several disorders of the lower bowel, but it may also occur as a distinct affection, as independently of any other disease of the part, being due to a peculiar hyperæsthesia of the skin. Prurigo ani is caused by worms in the lower part of the rectum, and by congestion of the hæmorrhoidal veins. In women it is consequent on affections of the womb. Patients suffer most after taking stimulating drinks, and during warm weather and when heated in bed. The itching is extremely teasing and annoying, especially at night, when it sometimes keeps the patient awake for hours. Rubbing the part to arrest the irritation only aggravates the mischief afterwards, yet few persons have sufficient self-control to prevent their seeking temporary relief by friction, and some, though capable of restraining themselves whilst awake, fret the part unconsciously during sleep. The friction thus resorted to excoriates the skin at the margin of the anus, so that in chronic cases the skin becomes dry, harsh, and leathery, cracks from slight causes, and ulcers and fissures are produced, which are but little disposed to heal. In most instances this complaint, after proving troublesome for an hour or two at night and in the day after stimulants, ceases, and the patient has long intervals of rest and ease. But in the worst forms of the malady, the torment is most distressing. It lasts throughout the night, so that the patients get little but broken sleep, and after a time the general health suffers seriously, and life is rendered truly miserable. In some of the cases which have fallen under my notice, I could discover no local cause whatever to account for the prurigo. It seemed to be purely an affection of the nerves of the part. The patients are generally healthy. One gentleman who had been subject to it for years, found that it was connected with his state of mind. When much engaged and prosperous in business, he suffered little from it. He was sometimes free for a whole month, and then became troubled for many nights in succession. In cases of this kind the complaint, after proving troublesome for years, has been observed to subside as age advances.

In prurigo ani the habits of living should be regulated. The patient should sleep on a mattress, and be as lightly covered as is consistent with comfort, cold bathing or sponging should be daily resorted to, and sufficient exercise taken in the open air. Stimulants and hot condiments must be strictly avoided. The actions of the bowels are to be regulated if necessary by medicine, and after each evacuation the parts should be cleansed with soap and water. Every effort should be made to avoid friction, and the patient should be assured that if he yields to his inclinations, his complaint will be rendered worse and more difficult of cure. In all cases, the condition producing this troublesome symptom must be the chief object of attention, such as worms, congestion, &c., but there are certain remedies which are specially adapted to relieve the irritation. The itching attendant on piles may be arrested by smearing the anus with some mercurial ointment, as the dilute citrine, or one containing the gray oxide of mercury, or by lodging in the parts a piece of cotton-wool soaked in a lotion of oxide of zinc. Lotions of carbonate of bismuth and glycerine, of borax and morphia, or of carbolic acid, are often efficacious in this complaint. The application to the anus of strong solution of nitrate of silver (gr. xx— ℥ j) with a camel's hair brush once daily often gives relief, especially in cases where the skin is made rough and sore by rubbing. In some cases great benefit has been derived from chloroform ointment. It produces a smarting sensation when first applied, but this is soon followed by ease. In persons of weak constitution benefit has resulted from full doses of quinine, and in certain cases liquor arsenicalis with steel has helped to relieve the irritation. I have sometimes found it necessary in severe cases to order suppositories of morphia at bed-time. The complaint is often very obstinate, and much perseverance is required on the part of the practitioner, and also of the patient, to effect a cure.

INTESTINAL WORMS.

By W. H. RANSOM, M.D., F.R.S.

INTRODUCTORY REMARKS.—No definition of the disease, such as stands at the head of each article in this volume, is requisite or appropriate in treating, from the point of view of the practical physician, of the parasitic worms which inhabit the human alimentary canal. But it may be desirable briefly to indicate the general scope or plan of this article, as well as the limits within which it will be restrained.

In most diseases, as for instance in the exanthemata, a brief summary of the more constant phenomena may serve at once as a definition and means of diagnosis; but, as the external agents or exciting causes of those phenomena escape our search, the etiology of such diseases is little more than an investigation of the conditions favorable to their occurrence, with speculations upon the nature of the exciting cause: while the pathology is limited to a consideration of the relations existing among the phenomena observed during life or after death, and between these and the favoring conditions.

But in the medical study of parasites the whole question of "the changes from a condition of health" is viewed from quite another standpoint. Here we can begin with the exciting cause, which we can isolate, compare, experiment upon, and learn the natural history of, before we study its effects. The extension of knowledge may possibly hereafter enable us so to approach the study of cholera or scarlet fever.

In this article the order thus indicated will be followed; the names and zoological position of the worms found in human intestines being first stated, the more important species will be described and their life histories traced, with only so much of detail as may be required for the purposes of the medical practitioner. Afterwards the changes of function or structure which they produce, the conditions which favor their occurrence, the mutual relations of the observed phenomena, the methods of detecting, expelling, and avoiding these pests, will be treated of.

Those parasitic animals belonging to the Gregarinida and Infusoria, as well as the accidental or occasional but not truly parasitic inhabitants of our intestines, such as insect larvæ, will be excluded from consideration here on account of their at present comparative insignificance clinically. The *Trichina spiralis* will also be passed over, because, although it attains its state of sexual maturity in human intestines, its importance to the physician depends upon the habit which its larvæ have of perforating the tissues and becoming encysted in the muscles. Moreover the very great importance which has recently attached to this worm justifies the devotion to it of a separate article.

It is difficult, if not impossible, adequately to appreciate the relation of intestinal worms to their bearers without including in the investigation the lower animals. To do so here would, however, be foreign to the design of this work, and the reader who seeks for fuller information on this subject will do well to consult the works of Kückenmeister, Von Siebold, Davaine, Cobbold, and especially of Leuckart. I may however draw attention to two prominent results of the comparative study of Entozoa. They are so widely diffused that scarcely any species of animal is known which is not, at least sometimes, infested by them; and notwithstanding the fact that they can, and do, often injuriously and even fatally influence the animals they infest, yet in the majority of cases the observer is struck with the apparently trivial inconveniences they produce.

HISTORY.—The intestinal worms, or some of them, have been known from very early times. Hippocrates mentions the tape-worm, and Aristotle described in addition the round-worm and the seat-worm. During the classical and middle ages the doctrine of spontaneous generation held general sway, and was thought to afford a satisfactory explanation of the then known facts as to the occurrence of Entozoa. Although Swammerdam[1] and Redi[2] shook the foundations of this doctrine in its application to insects and their larvæ, they did not venture to apply their views to the Entozoa. The first great step towards sounder views was made by Pallas,[3] who taught that Entozoa, like other animals, sprang from similar parents, and were propagated by means of eggs which were transmitted from one host to another. But in the absence of direct evidence these opinions were for a time borne down by the authority especially of Rudolphi[4] and Bremser,[5] who reverted to the doctrine of spontaneous generation. Soon, however, the progress of biological science, aided by improved means of research, and directed into new channels, broke down this doctrine at once and for all time, at least in its application to intestinal worms; and the researches of Mehlis (1831),[6] Von Siebold (1835),[7] and Eschricht (1837),[8] confirmed the main proposition of Pallas, and justified the conclusion of Eschricht, that Entozoa during their reproduction generally undergo a metamorphosis and a migration. Then followed the brilliant discovery of alternation of generations by Steenstrup (1842),[9] the researches of Von Siebold (1848),[10] and Van Beneden (1850),[11] and the true life history of the *Trematoda* and *Cestoda* was understood. It remained to furnish direct proofs of the correctness of the new views, and these were given by Kückenmeister (1852),[12] who fed carnivora on flesh containing *Cysticerci* and produced tape-worms, and by feeding herbivora with ova of Tæniæ produced *Cysticerci*. Many other zealous and able investigators in this country, as well as in France and Germany, have confirmed his results, and otherwise extended our knowledge of the intestinal worms. Prominent

---

[1] Bibel der Natur. Ausdem Holl. übersetzt. 1752.
[2] Esperience intorne agl' Insetti. 1712.
[3] Neue Nord. Beiträge. 1781.
[4] Entozoor, hist. Natur, vol. 1.   1808.
[5] Ueber lebende Würmer im lebenden Menschen.   1819.
[6] Oken's Isis. 1831.
[7] Archiv für Naturgeschichte.   1835.
[8] Nova Acta Academ. C. L., vol. xix.   1837.
[9] Ueber den Generationswechsel.   1842.
[10] Jahresbericht im Archiv für Naturgeschichte.   1848.
[11] Les Vers Cestoides.   1850.
[12] Prager Vierteljahrschrift.   1852.

among these stand the names of Haubner, Leuckart,[1] Dujardin,[2] Davaine,[3] and Cobbold.

The opinions of medical men as to the clinical importance of intestinal worms have varied with the changes of biological theory, usually lagging somewhat behind, but depending mainly upon it. So long as the doctrine of spontaneous generation in any of its forms was believed to account for the presence of Entozoa a mysterious dread of their power for evil prevailed, and evidenced itself by the multitude of grave diseases attributed to them. Indeed few maladies afflict humanity which were not sometimes attributed to intestinal worms, even by prominent men in their day.

This was due not alone to the common tendency to magnify the unknown, but also to the uncertainties of diagnosis, the absence of a pathological anatomy, and the frequency with which worms were observed to pass away in the course of serious diseases, the subsequent recovery from which being imputed to their escape.

In the latter half of the eighteenth century an extreme reaction took place among those who gave themselves specially to the study of Entozoa, so that it was maintained that they were beneficial to their hosts, or at most only very rarely and accidentally injurious.

The physicians as a rule, however, still clung to the older views, and in doubtful cases found a ready and satisfactory explanation of the symptoms in the assumption of an irritation by imaginary worms. Even Rudolphi and Bremser, while opposed to the prevalent medical opinion, sought to explain the actual symptoms which attended the presence of worms in the intestines by the hypothesis of a pre-existing diathetic state (Helminthiasis), which they believed to be a necessary condition of the spontaneous development of worms. Only in the present generation have sound views on this subject prevailed, and only since the discoveries of Kückenmeister and his followers has a satisfactory knowledge of the life history of human intestinal worms enabled the physician to appreciate their true importance in medicine, to ascertain their presence with certainty, and in most instances to point out how they may be avoided.

Out of at least thirty-one Entozoa which are at present known to inhabit our bodies, thirteen infest the alimentary canal. Of these seven belong to the order *Cestoda :—*

1. *Tænia solium*, Linnæus.
2. *Tænia medio-canellata*, Kückenmeister.
3. *Tænia nana*, Von Siebold.

4. *Tænia flavo-punctata*, Weinland.
5. *Tænia elliptica*, Batsch.
6. *Bothriocephalus latus*, Bremser.
7. *Bothriocephalus cordatus*, Leuckart.

And six to the order *Nematoda :—*

8. *Ascaris lumbricoides*, Linnæus.
9. *Ascaris mystax*, Rudolphi.
10. *Oxyuris vermicularis*, Bremser.

11. *Dochmius duodenalis*, Leuckart.
12. *Trichocephalus dispar*, Rudolphi.
13. *Trichina spiralis*, Owen.[4]

ORDER *CESTODA.*

Parenchymatous worms, without mouth or alimentary canal, with a so-called water-vascular system. They develop by budding from a pear-shaped larval form (scolex) to a long, jointed, tape-shaped colony of individuals (strobila). In their reproduction they suffer an alternation of

[1] Die menschlichen Parasiten, &c. 1802-68.
[2] Histoire Naturelle des Helminthes. 1845.
[3] Traité des Entozoaires. 1860.
[4] See article *Trichina spiralis.*

generations. The individual members of the colony (proglottides), or sexually ripe animals, increase in size and complexity of structure, although otherwise resembling each other, the further they are removed from the head, near to which a continuous formation of new joints takes place by budding. The head, which is the same in the adult as in the larval form, is furnished with two or four suckers, and commonly also with a coronet of hooklets, which serves for attachment. They infest in their adult state the alimentary canal of vertebrate animals only. The ovum yields a globular embryo furnished with three pairs of hooklets, and develops into the Scolex (*Cysticercus*) in the tissues or in parenchymatous organs, usually of food animals, and is thence passively transferred with the food into the intestine of its definitive bearer, where it assumes the adult form.

TÆNIA SOLIÙM (Linnæus)

Was at one time believed to be "the common tape-worm of man," but it is now known that at least one other species is included in that expression.

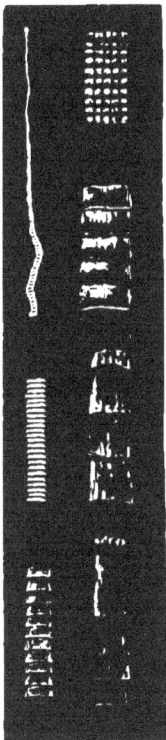

FIG. 2.—Head of *T. soli-um*. (Davaine.)

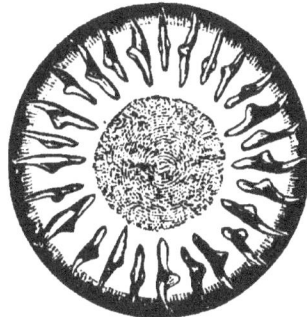

FIG. 3.—Coronet of hooks, magnified. (Leuckart.)

FIG. 1.—*Tænia solium* natural size. (Davaine.)

FIG. 4.—Separate hooks, more highly magnified. (Leuckart.)

Description.—The adult worm (Strobila, Fig. 1) commonly attains a length of from 7 to 10 feet,[1] but is often much longer. The number of

---

[1] This is Leuckart's measurement. but there is a wide divergence among authorities on this point. Davaine makes the common length from 20 to 26 feet.

joints increases with the length; a worm measuring 7 ft. 6 in., counted by Leuckart, had 749 joints. The head (Fig. 2) has a somewhat globular form, measures about $\frac{1}{15}$ in. to $\frac{1}{15}$ in., is marked anteriorly by a moderately prominent rostellum, bearing a crown of about twenty-six hooks, and by four projecting suckers.

The threadlike neck is nearly an inch in length, and to the naked eye is not distinctly jointed; it passes gradually into a jointed, continually widening band of a whitish color, of which the earlier segments are so much shorter than broad that one-half of the whole are found in the anterior ninth of the chain. Slowly the joints increase in length more than in breadth, so that they assume a square form about the end of the anterior third. Mature joints, *Proglottides* or *Cucurbitina* (Fig. 5), measure about $\frac{1}{2}$ in. in length and $\frac{1}{4}$ in. in breadth, being now longer than broad. They are flat and thin, with a quadrangular outline, are furnished with a longitudinally placed tubular uterus, having seven to ten branches on each side, within which are seen developing ova. Male and female organs of generation are present in the same joint, and open by a common aperture near the centre of one or other border, now right, now left. The sexual organs are already distinctly visible in the joints at one-ninth of the whole length from the head, the ova are impregnated about another ninth lower down the chain, and soon afterwards the eggs enter the uterus.

The water-vascular system consists of a single longitudinal canal at each border, and one transverse, near the posterior edge; it is continuous from one segment to another throughout the chain. The cystic worm known as *Cysticercus cellulosœ* is the larval form, or *Scolex* (Figs. 6, 7); it is commonly found in the flesh of pigs, but occasionally also in other animals, and even in man: the adult colony has only been found in man. The eggs (Fig. 8) are globular in form, measure when free about $\frac{1}{170}$ in.,

FIG. 5.— Ripe joints of *T. solium*, magnified. (Leuckart.)

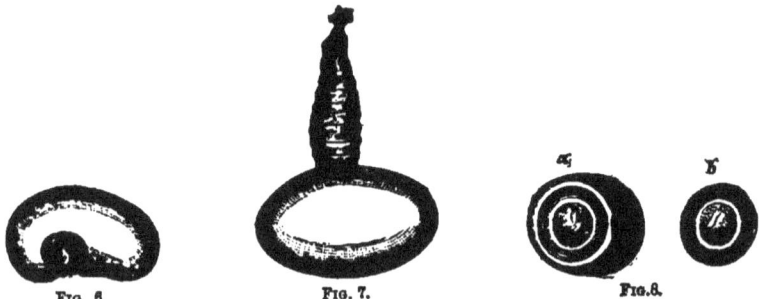

FIG. 6.

FIG. 7.

FIG. 8.

FIG. 6.—*Cysticercus cellulosœ*, natural size and position. (Leuckart.)
FIG. 7.—*Cysticercus cellulosœ*, magnified. Head and neck protruded. (Leuckart.)
FIG. 8—Ripe ova of *T. solium*. *a* with outer capsule as seen in uterus: *b* free, as found in fœces. (Leuckart.)

have a thick firm shell of a brownish color, radially and concentrically striated, and when taken from the uterus often an outer capsule with a more oval outline (Fig. 8, *a*). The contained embryo is globular, and furnished with three pairs of hooklets. A moderate-sized tape-worm has been calculated to contain about 5,000,000 of ripe ova.

Life History.—The normal habitat of *T. solium* is the small intestine of man: Kückenmeister has seen it while yet alive firmly attached by suckers and coronets to the mucous membrane. Formerly it was believed that it was always solitary, and this error perhaps explains the statements made by the older authorities of the occurrence of worms of enormous length. It is now known that although commonly one, two, or three are found together, yet various numbers, up to forty at least, may be present.

From the lowest end of the band—which hangs a variable distance down the intestines, and may reach the colon—ripe joints spontaneously separate and escape with the fæces, either singly or united into short lengths. Frequently, also, ripe ova escape by rupture from the joints into the intestine and mingle with its contents. The free joints in moist and warm situations move about for a time, and by this and other accidental agencies the ova are widely disseminated; doubtless the vast majority fail to find suitable conditions for their development, and therefore die; but a small proportion of joints or ova are taken with the food into the stomach of a pig, or much more rarely into that of a man; where, after digestion and rupture of the shell, the embryo (*pro-Scolex*) escapes, and by diligent use of its armature perforates the tissues of its involuntary host, and ultimately settles down in some, to it, suitable locality, generally the cellular tissue of the muscles, but sometimes the liver or the brain. The embryo there remains quiet, in some organs is encysted, undergoes a metamorphosis, and becomes the well-known *Cysticercus cellulosæ* of measly pork (Figs. 6, 7). As usually found, it has the head and neck inverted, and its characters are difficult to observe, but when everted is seen to have a head and neck like that of *T. solium*, with a vesicular caudal appendage. This metamorphosis requires about two months and a half for its completion; afterwards the *Cysticerci* remain without further change, but capable of further development, if the proper conditions are supplied, for a period not yet certainly known, but which has been estimated at from three to six years.

When the flesh of pigs so infested is eaten raw or imperfectly cooked, the *Cysticercus* is partly digested in the stomach, so as to lose its vesicular annex; it then passes into the small intestine, and, attaching itself, becomes developed in about three to three and a half months into the adult form already described, which may continue to infest its bearer for ten, or even, it is said, thirty-five years. It would take too much space here to recount the evidence upon which this summary statement rests; but it may be said in brief that Kückenmeister, Leuckart, and others have, notwithstanding some opposing statements, placed it beyond reasonable doubt by a carefully devised and executed series of experiments, in which pigs have been infected with *Cysticercus cellulosæ* by eating ripe joints of *Tænia solium*, and men have been infected with tape-worm by eating measly pork.

This biography of *T. solium* illustrates that of other parasites of the same group, and the study of each has thrown light upon the others: for this reason, and to show the relation between the food of animals and their parasites, the following short list may be permitted a place here:—

Cysticercus fasciolaria in the mouse is the larval form of *Tænia crassicolles* in the cat.

Cysticercus pisiformis in the rabbit is the larval form of *Tænia serrata* in the dog.

Cysticercus tenuicollis in sheep, oxen, &c., is the larval form of *Tænia marginata* in the dog.

Cænurus cerebralis in sheep is the larval form of *Tænia cænurus* in the dog.

Cysticercus tæniæ medio-canellatæ in the ox is the larval form of *Tænia medio-canellata* in man.

Symptoms.—There can be no question that a large proportion of persons infested with this tape-worm are unconscious of any departure from the state of perfect health, but there is as little doubt that in some instances functional derangements occur which are referable to the local irritation it produces. In a much smaller number of cases and under exceptional conditions, even structural changes are produced by it.

The functional derangements belong to two groups. (*a*) Those excited in the part irritated, and its immediate neighborhood. Such are, various uncomfortable sensations in the abdomen, pains resembling colic, sometimes felt when the stomach is empty, at others after certain articles of food, variable appetite, now excessive, now failing entirely, slight diarrhœa, or constipation, &c. (*b*) Those of reflex origin. These are itching of the nose or anus, headache, giddiness, ocular spectra, tinnitus aurium, palpitation, cardialgia, increased flow of saliva, nausea, lassitude, pains in the limbs, and an uncertain flow of spirits. In women, disordered menstruation, spasmodic and convulsive movements, hysterical fits, and even epileptic and maniacal attacks, have been said to be due to their irritation. In long-continued cases, Kückenmeister thinks wasting has been produced. This somewhat grave list of symptoms contains little or nothing that is characteristic of the nature of the irritative cause, and must be received with some caution, on two grounds: one, that patients not unfrequently exaggerate their sensations when they either have had, or have suspected themselves to have had, worms of any kind; and the other, that the symptoms enumerated have in great part been collected and handed down to us from earlier times, when medical men, not yet familiar with the results of comparative helminthology, shared, to some extent, the common mysterious dread of Entozoa, and too hastily attributed the observed phenomena to the influence of worms, which were indeed present, but not necessarily acting as exciting causes. In support of this assertion, it is sufficient to recall the fact that many healthy persons are infested with tape-worms and present no symptoms; and also, that many persons suffering from various diseases have tape-worms, and these more than other persons are apt to expel them, and thus mislead.

It may, nevertheless, be readily granted that those who have a delicate or irritable mucous lining to their intestines, or who are of a nervous temperament, and abnormally liable to reflex excitement, do suffer some, perhaps many, of the symptoms here recounted, and that in stronger persons the same may happen if the worms are very numerous. But it is worth remembering, that paroxysmal maladies, such as convulsions, mania, &c., are peculiarly liable to give rise to errors in reasoning as to their causes, so that very rarely could it be affirmed that they were caused by a tape-worm when their cessation coincided in time with its expulsion.

In some cases, proportionally few in number, when abscesses have formed in connection with an obstruction of the intestine, a tape-worm has escaped from the opening, and may have been partly, or perhaps solely, the cause of such obstruction and abscess.

There is another fortunately rare, but grave, consequence of the presence of a tape-worm; it may give rise to the development of the *Cysticercus cellulosæ* in the tissues or organs of its bearer, and thus even destroy life. This may conceivably take place when, as a consequence of violent vomiting, some of the ripe joints are carried up into the stomach, where the digestive fluids might set the embryo free; or in the case of children or dirty people, by conveying the escaped segments or free ova, upon the hands or with the food into the mouth, and thence into the stomach.

Diagnosis.—When a patient presents such a conjunction of symptoms as, in the absence of other indications, excites a suspicion of tape-worm, its presence can only be ascertained by an inspection of the stools. The ripe segments (Fig. 5) or the ova (Fig. 8, *b*) will with a little care almost certainly be found in the fæces, and from them the species may be determined with sufficient exactitude for the requirements of the physician.

Etiology.—The exciting cause of the disease is manifestly the worm, a foreign irritating body in the intestine. The favoring conditions are the adult age, possibly the female sex, certainly some occupations, such as those of the cook or the butcher, the habit of eating raw or underdone pork, ham, sausages, &c., and a residence in Europe, India, Algeria, North America, and probably wherever the pig is domesticated.

Pathology.—Leuckart has shown by observations on the dog, that local congestions of the mucous membrane, separation of the epithelium, and even minute superficial sores, may result directly from the activity of a tape-worm. If it be admitted that *T. solium* may cause similar local changes in man, there is no difficulty in connecting the deranged functions of the alimentary canal with the worm as their cause, if we grant either an exceptional delicacy of the bearer, or an unusual number of worms. The remote functional disorders present no more difficulty, if pre-existing abnormal proclivity to reflex movements be granted.

Treatment.—The indications for treatment follow in the clearest manner from the foregoing. The worm as exciting cause must be got rid of, and the effects then commonly subside; but should they persist for a time, they can be successfully met by suitable diet and the treatment for irritation of the intestines.

An immense number of substances have, at various times, enjoyed a reputation for the possession of anthelmintic powers, too often without any accurate distinction of the kind of worm, so that with the rise of a more accurate diagnosis, as well as, perhaps, of a more critical spirit in modern times, the number of accepted remedies for tape-worm has rather diminished, and a general demand has arisen for a re-examination of the claims of most of the reputed agents.

THE MALE SHIELD-FERN (*Aspidium filix mas*) is perhaps the oldest and most widely known vermifuge, and of late has grown into much favor, especially in this country.

The dose is from 60 to 100 grs. of the powder of the dried rhizome, or from ℨj. to ℨij. of the liquid extract, given upon an empty stomach, preceded and sometimes followed by a purgative. It has been said to act by killing the worm; it certainly has a violent and irritating action upon the lining membrane of the stomach or bowels, often causing vomiting, and in large doses purging, with slimy and even bloody stools.

THE BARK OF THE POMEGRANATE ROOT (*Punica granatum*), also an ancient and extensively used remedy, is recommended by Bamberger as the best and least disagreeable in its action of all the remedies for the expulsion of tape-worm. He insists upon its being used fresh, and considers the old and dry bark almost inert. He prepares the patient by spare diet and aperient medicines, and then gives a pint of a decoction much like that of the British Pharmacopœia (equal to 2 oz. of bark) in three doses, at short intervals, early in the morning. Kückenmeister uses a still stronger decoction, and gives a quantity equal to 4 oz. of the pomegranate bark, with 20 grains of the ethereal extract of male fern added. The German authorities generally employ powerful, not to say violent measures, for the expulsion of tape-worm, but how far this may be due to the greater resistance which some species present is unfortunately not yet certain.

Kousso—the flowers and tops of *Brayera anthelmintica.*—In doses of ¼ to ½ oz. or more it is a quick and good anthelmintic, much used in Abyssinia for the species of tape-worm there prevalent. It is not much used in Europe, perhaps on account of its cost, of the difficulty of obtaining it, and of the inconvenient form in which it is usually administered.

Kamala, from the fruit of the *Rotleria tinctoria, oil of turpentine,* and a number of other agents, have been recommended, but it is not desirable to notice them here. Some rare instances occur in practice, in which treatment by any or all of the above-mentioned drugs fails to expel the worm so as to prevent its recurrence, which takes place probably whenever the head and neck remain attached. Some cases indeed are recorded in which even the expulsion of the greater part of the band is not effected; and this not only when moderate doses have been used, but even after elaborate preparation, vigorous treatment, and free subsequent purgation such as Wawruch and other German authorities have advised. No very satisfactory explanation can be offered of this singular power of resistance occasionally met with; but in presence of the admitted failure of violent irritating remedies, it would seem prudent in such cases to continue moderate doses of male fern or pomegranate for longer periods of time, in conjunction with rigid prophylactic rules, to prevent the possibility of reinfection.

Prevention.—Each person can secure himself against *Tænia solium* by eating only such pork, ham, sausages, &c., as are well cooked; but the public health is not so easily cared for; it requires that pigs infested with measles should not be sold as food, and doubtless fewer pigs would suffer from measles were greater care taken to remove or destroy human excrement.

The *Cysticercus cellulosæ* when a human parasite, is treated of in another part of this work.

TÆNIA MEDIO-CANELLATA (Kückenmeister).

Description.—This worm was formerly held to be an unarmed variety of *T. solium,* but Kückenmeister and Leuckart have recently established its specific distinctness both by observation and experiment. It has a general resemblance to, but is larger and firmer in texture than, *T. solium;* not only does the whole band (Strobila, Fig. 9) commonly attain a greater length, but the segments are more numerous, and larger in all their dimensions. The unripe ones are broader than long, the ripe ones longer than broad. The contained uterus (Fig. 10) is more finely divided than in *T. solium,* having from 20 to 35 branches on each side. The common sexual aperture is placed alternately on either border, nearer to the posterior margin than in *T. solium.* The head is large (Fig. 11), measuring about ₁₁₂ in. (Davaine); has neither rostellum nor coronet of hooks; is furnished with four very powerful and prominent suckers; and, according

11

to Leuckart, a fifth smaller one in the usual position of the rostellum (Fig. 15). Kückenmeister also figures a central canal connected with the water vascular system.

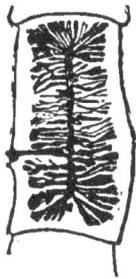

FIG. 12.—Ovum of *T. me-dio-canellata.* (Davaine.)

FIG. 10.—Ripe joint of *T. medio-canellata.* (Leuckart.)

FIG. 11.—Head of *T. medio-canellata,* magnified. (Davaine.)

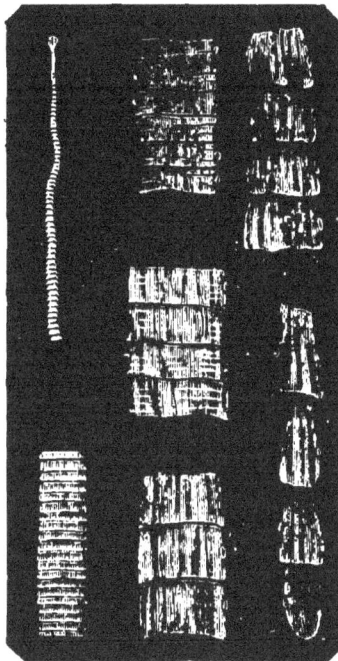

FIG. 13.—*Cysticercus T. medio-canellatæ,* natural size and position. (Leuckart.)

FIG. 9.—*Tænia medio-canellata,* natural size. (Davaine.)

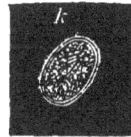

The eggs (Fig. 12) resemble those of *T. solium,* except that they are more oval in outline: they measure about $\frac{1}{700}$ in. by $\frac{1}{850}$ in.

The larval form, or *Cysticercus tæniæ medio-canellatæ* (Figs. 13, 14, and 15), infests the flesh and organs of the ox, a fact which at once points out the chief difference between its life history and that of *T. solium.* *T. medio-canellata* abounds in Abyssinia and South Africa, and is also common in Europe: it was, until the recent researches of Dr. Cobbold, thought to be more common in continental states than in this country; but it is now known to occur almost, if not quite, as frequently amongst us as *T. solium* does.

FIG. 14.—*Cysticercus T. medio-canellatæ.* Head everted. Magnified. (Leuckart.)

Nothwithstanding its being unarmed, the great strength of its suckers enables the head to hold on with even greater tenacity than the *T. solium,* so that it is more difficult to expel, and it is believed to excite more marked symptoms; but as the larval form does not, so far as is at present known, infest man, it is less dangerous to life.

The terminal joints separate spontaneously from the parent chain, and often creep out of the anus irrespective of the passage of fæces; as a rule having first permitted at least a portion of their contained ova to escape by rupture into the intestine.

So far as is at present known, its treatment is the same as that for *T. solium*, and its prevention consists in the avoidance of raw or underdone beef.

The three following *Tæniæ* are placed by Leuckart in a separate group, of which the larvæ are distinguished by having comparatively small caudal vesicles, and are met with only in cold-blooded, generally invertebrate animals. Those occurring in man are minute, and have been so rarely met with, at least in Europe, as to be of comparatively little clinical importance.

TÆNIA NANA (Von Siebold) Is scarcely an inch long, and about $\frac{1}{50}$ in. wide at its broadest part. Head globular, with an oval rostellum bearing a single row of 22 to 24 very minute hooks, and four rounded suckers. Eggs globular, $\frac{1}{840}$ in. Found by Bilharz in great numbers in the duodenum of natives of Egypt. Its migrations and metamorphoses are unknown.

TÆNIA FLAVO-PUNCTATA (Weinland).

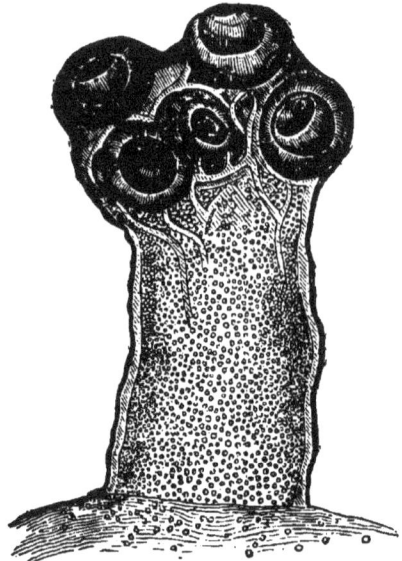

FIG. 15.—Head of *Cysticercus T. medio-canellatæ*, more highly magnified, showing central sucker. (Leuckart.)

The adult attains a foot in length. The joints of the anterior half of the chain are marked by a distinct yellow spot, the receptaculum semini, which is absent in the following segments. Head unknown. The egg measures $\frac{1}{125}$ in. Met with but once in a healthy infant in North America. Life history unknown.

TÆNIA ELLIPTICA (Batsch).

The adult worm attains a length of 6 in. to 8 in., head very minute, measuring $\frac{1}{80}$ in., rostellum cylindrical, furnished with three or four rows of hooklets. Terminal segments three or four times as long as broad. Sexual apertures double, one on each margin of segment. · Eggs measure $\frac{1}{500}$ in. It infests normally the intestine of the cat, and only very exceptionally has been found in man. Its life history is unknown.

FIG. 16. — Ovum of *Tænia* of uncertain species.

Here I venture to add an abstract of a case (*Med. Times and Gazette*, p. 598, 1856) which suggests the possible addition of still another species of *Tænia* to the above list. A girl aged nine years, suffering from disordered digestion and impared nutrition, passed with the fæces for more than fifteen months consecutively numerous oval ova (Fig. 16), measuring about $\frac{1}{500}$ in. by $\frac{1}{500}$ in. and containing a globular embryo, furnished with three pairs of hooklets. These eggs differed so much from those of any other tape-worm then known to me, that I referred them to an undescribed species of *Tœnia;* but whether this may ultimately prove to be correct or not, the view receives some support from the fact, that although the girl during the whole of that time was under observation as a hospital patient, was treated vigorously and repeatedly with male fern, kousso, pomegranate bark, turpentine, and various cathartics, and the stools carefully watched, yet no tape-worm joints were ever found, although the ova continued to be expelled in undiminished numbers. It is very difficult to suppose that the child harbored a *T. medio-canellata* which, although sexually mature, passed no joints, yet this is the only *Tœnia* known to me of which the ova have even a passing resemblance to those found in this case. It seems more probable that the tape-worm was one which normally expels its ova without casting off joints of such dimensions, or in such a condition, as to be recognizable in the stools on a careful search.

In this case the functional disorder subsided shortly after treatment began, but as the ova continued to escape, it could not have been caused by the parasite or parasites. Ultimately the patient ceased to attend, but to the last her fæces contained the same ova.

The two remaining tape-worms of man belong to the family *Bothriocephalidæ*, of which the adult forms infest chiefly cold-blooded vertebrate animals.

BOTHRIOCEPHALUS LATUS (Bremser).

Description.—This is the largest tape-worm known to inhabit man; it commonly reaches a length of 17 to 26 feet, and sometimes 60 feet or more (Fig. 17). The head (Fig. 18) is unarmed, oblong, or club-shaped, it measures $\frac{1}{10}$ in. in length by $\frac{1}{20}$ in. in breadth, has a deeply-grooved longitudinal sucker on each side, and passes gradually into a short thread-like neck. The joints are broader than long, the widest being $\frac{1}{4}$ in. in length, by $\frac{1}{2}$ in. or even more, in breadth; towards the posterior end of the chain they increase in length and diminish in breadth, assuming thus a more square form; they are thicker in the middle than at their margins, from the presence there of the sexual organs, which form a rosette-shaped patch in the centre of which the sexual apertures are placed. The eggs (Fig. 19) are oval, $\frac{1}{310}$ in. by $\frac{1}{410}$ in.; have a firm, brownish, structureless shell, with an operculum at one end. While yet within the uterus they present no trace of embryo in their interior.

Life History.—The ova escape by rupture of the ripe joints, and probably in part also through the oviduct, into the intestine before the joints separate; these are expelled with the stools at rather long intervals; not singly, as is often the case with *T. solium* and *T. medio-canellata*, but in short chains of 2 to 4 ft. in length. The ovum after a prolonged sojourn in water develops a ciliated embryo, which escapes through the aperture in the shell by forcing open the lid, and is furnished with three pairs of hooklets. On analogical grounds it is very probable that it then enters into the body of some aquatic animal, possibly a fish, but probably a mollusc, and then assumes the larval form, which is at present unknown. The intermediate bearer is probably eaten by man, and the larva assumes the adult form in his intestine. *B. latus* usually occurs several together; it has a somewhat limited geographical distribution, not having been found beyond the limits of Europe; in some countries of which only is it indigenous. It is common in the western cantons of Switzerland, North-western Russia, Sweden, Poland, Holland, Belgium, and Eastern Prussia; it is less often met with in other parts of Germany, and has occasionally been imported into Britain. Low-lying damp regions near the borders of seas and lakes are those in which it is most often abundant. It is found in persons of all ages and sexes, and in those countries where it is most frequent, even children at the breast are not free from it.

The *Symptoms* do not differ from those caused by the species of *Tæniæ*. Its presence may be detected by an examination of the stools. It may be expelled by the same drugs as are employed in the treatment of other tape-worms, and it is said to be less difficult to dislodge, perhaps on account of the feeble development of its suckers. No precise knowl-

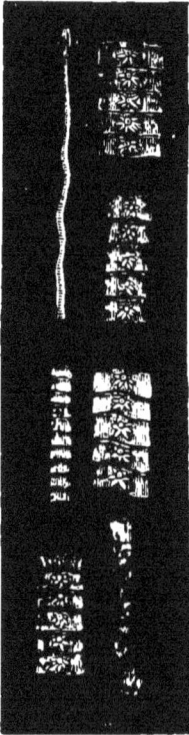

FIG. 18.—Head of *Bothrio-cephalus latus*, magnified. (Da-vaine.)

FIG. 17.—*Bothriocephalus latus*, natural size. (Davaine.)

a.   *b*

FIG. 19.—Ova of *Bothrio-cephalus latus:* *a* with contained yolk; *b* empty shell. (Leuck-art.)

edge has yet been attained of the measures to be taken to avoid it; but the general rule of carefully cooking all foods and of drinking only pure water would be likely to succeed even in those countries where it most abounds.

BOTHRIOCEPHALUS CORDATUS (Leuckart).

A recently discovered, much smaller worm, found only in North Greenland in men and dogs. It is known by its caudate head and the absence of a neck.

ORDER *NEMATODA.*

Elongated, slender, often thread-like worms, not distinctly jointed, or provided with appendages; with a separated alimentary canal, a terminal mouth, an anus (*Gordius* excepted) near the cau- dal extremity, opening on the ventral aspect. The integument is marked by two lateral longitudinal bands, and often by a dorsal and a ventral one; in the former are embedded the nerves with their ganglia, and the excretory tubes which open in the surface about the level of the pharynx. The fe- male aperture is placed near the central region of the body, that of the male near the anus, and con- joined with it; it is furnished with retractile spiculæ, usually two or more. The male is smaller than the female. The development is direct and the meta- morphosis inconspicuous; so that the embryo has the general aspect of a nematode worm. The order is rich in species, and furnishes as many parasites as all the other Helminthoids put together. They infest invertebrata as well as vertebrata, and no organs escape their invasion.

ASCARIS LUMBRICOIDES (Linnæus).   Common round-worm.

Description.—A large nematode worm, during life of a reddish or brownish tinge, and of a firm, elastic texture (Fig. 20). The female reaches 15 in. in length by ⅛ in. to ¼ in. in breadth; and the male 10 in. by ⅛ in. (Leuckart).[1] The cylindrical body, covered by a cuticular layer and marked by fine transverse rugæ, tapers towards both ends, but more rapidly towards the head; in which is placed the terminal mouth, surrounded by three nearly equal prominent muscular and tactile lips (Fig. 21), each nearly as high as broad, and marked off at its base by a distinct groove. The inner surface of each lip is beset with about two hundred very mi- nute microscopic teeth. The triangular mouth con- ducts to a muscular œsophagus, and this to a sim- ple, almost straight intestine, without distinction of stomach. The lateral longitudinal bands, much more distinct than the median, divide the muscular mass into nearly equal areas, and give attachment to their fibres, as well as support the nerves and excretory tubes. The caudal extremity, short and conical, terminates in a point, and in the male curves strongly towards the ventral aspect, on which is seen the cloacal aperture with two often projecting spiculæ (Fig. 22). These are connected with a short, ejaculatory duct, which is continuous with a seminal vesicle, and a single long, tortuous, tubular testis; the whole male generative organ forming a tube eight times the length of the animal. The vulva in adult females opens about the junction

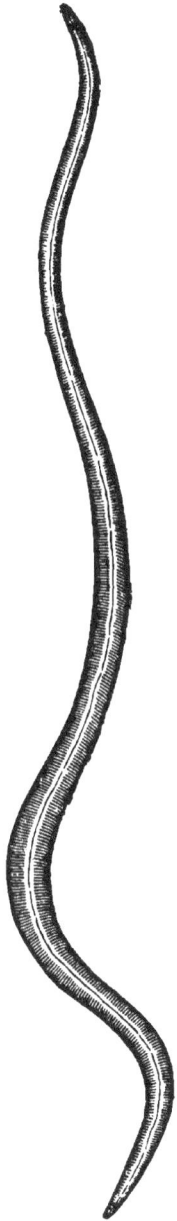

FIG. 20.—Adult female, Ascaris lumbricoides, natural size. (Leuckart.)

---

[1] These measurements exceed those given by Davaine.

of the anterior and middle third of the body, it conducts to a short vagina, this to a uterus, which soon divides into two long horns, directed backwards; each of these leads to a short oviduct, which serves also as a receptaculum seminis, and thence to a very long, tortuous, tapering ovary. The female generative tubes are eleven times the length of the adult animal. The ova are oval in form, and have a thick, firm, elastic, brownish shell, the surface of which is generally nodulated. No commencement of development is seen in their interior when deposited. They measure $\frac{1}{310}$ in. by $\frac{1}{410}$ in. (Fig. 23, a and b).

Life History.—So fertile is the round worm, that, at a moderate calculation, its yearly production of ova may be taken at 60,000,000, so that over 160,000 are daily discharged into the intestine of its bearer by one adult female worm. As, however, several are often present together, it is easy to understand that the stools of an infested person are so thickly strewn with the eggs as to make their discovery by the microscope an easy matter.

Although the migrations of the embryo of *Ascaris lumbricoides*, and the true history of its development, are not yet ascertained with sufficient exactitude, the labors of Schubert, Verloren, Davaine, Leuckart, and others, permit the following history to be given as an approximately correct statement of the facts. The ova deposited with the fæces very slowly develop an embryo in damp earth or water. The process may be complete in a month if artificial warmth be applied, but in nature it usually requires

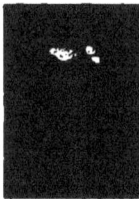

Fig. 21.   Fig. 22.   Fig. 23.

Fig. 21.—Head of *Ascaris lumbricoides*, magnified. (Davaine.)
Fig. 22.—Caudal extremity of male *A. lumbricoides*, magnified. (Leuckart.)
Fig. 23.—Ova of *A. lumbricoides*, from the stools; *a* recently deposited; *b* longer delayed in the stools. Shells tuberculated.

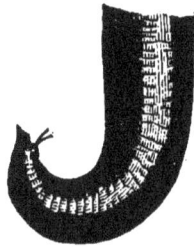

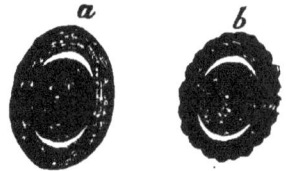

from five to eight months, and it may be delayed for a year or two by cold or dryness. Neither frost nor complete desiccation, however, kills the embryo, and the contained ova of dried females develop under suitable conditions. The ova do not normally hatch in a free state; Davaine has preserved them in water for five years without any visible change in the embryo, or spontaneous escape from the shell. In this stage the embryo has the general aspect of a nematode worm, with an alimentary canal, a commencing generative system, and a terminal boring, embryonic tooth. The next stage of their development is not known. Davaine maintains that the ova with their contained embryos are swallowed with impure water, and develop directly into the adult form if received into the intestine of a suitable bearer. But direct experiments do not support this view; dogs, rabbits, oxen, pigs, and men have been fed with large numbers of the ova of *A. lumbricoides* containing living embryos without any infection resulting. Similar experiments conducted upon horses, dogs,

and cats with the ova of their peculiar round-worms have had similar
negative results, and it seems indeed almost certain that infection does
not take place by a direct transference of the embryo-holding ova into the
alimentary canal of the definitive bearer.   It may be said also with some
confidence that the embryos do not escape from the ova to enjoy a free
existence for a time.   On analogical and other grounds it is a far more
probable view that the ovum is taken up in some way by an invertebrate
intermediate bearer, perhaps a worm, or the larva of an insect, and in it
the embryo passes through a necessary portion of its metamorphosis, and
then enters the stomach of its future host in some passive mode with food
or drink.

Ascaris lumbricoides infests also the pig[1] and the ox: it is found in
man all over the known world, but more abundantly in some countries than
in others.   In the Southern States of North America, especially among
the negroes, it attacks almost every one, young and old.   In the West
India Islands, Brazil, Finland, Greenland, in parts of Holland, Germany,
and France, it is also very frequently met with.   The rural population
suffer more than the dwellers in towns, and the inhabitants of low and
damp localities more than those who enjoy higher and dryer abodes.   The
poor, the young—excluding infants at the breast—the insane, and the
dirty, are peculiarly liable to be infested.   In certain regions it has occasionally prevailed so much for a time as to produce a kind of endemic
malady.

The round-worm normally inhabits the small intestine, and there is
some ground for the opinion that, unless a reinfection occurs, it escapes
after some months.   There can, however, be no doubt that it spontaneously wanders towards the external apertures under certain conditions
which are not well known, sometimes passing through the anus, the mouth,
the nose, often with severe purging, vomiting, or sneezing.   After death,
also, this migration is not uncommon, and is probably induced by a deficiency of food, or the presence of some conditions unsuitable for the wel-
fare of the worm; but whatever induces it, it results in placing the worm
occasionally in remote and singular localities, both during the lifetime and
after the death of the sufferer.   It creeps sometimes into the gall duct,
gall bladder, or hepatic duct, more rarely into the pancreatic duct, and
may give rise there to serious structural changes: it passes sometimes
through an ulcer or other abnormal opening in the intestinal wall, and
then is found after death in the peritoneal cavity, accompanied or not
with the signs of peritonitis, according as it may have migrated during
life or after death; it escapes sometimes with other intestinal contents
from abscesses or fistulæ in the abdominal walls, and appears, indeed, in
some such instances to have caused the local disease.   It has so marked a
tendency to creep into small apertures, that several instances are recorded
of its becoming fixed in the eyes of buttons and other similar small rings
which had been swallowed by the patient, and this habit has even suggested the swallowing of such rings to act as worm traps.   This migratory
instinct has occasionally led the round-worm along fistulous channels to
still more remote cavities or organs; for example, to the pleural sac, the
spleen, the kidney, the bladder, the muscles of the loin or neck, the spinal
cord, the lung, the glottis, the trachea, and the Eustachian tube.

In the more favored countries, usually from one to five worms are met
with together, but often many more are present; cases are recorded in

---

[1] Leuckart considers this species identical with *A. Suilla.*

which various numbers, from 200 to 2,500, have been expelled from one person within a few months, and 1,000 were found present together in the intestine of an idiot by Cruveilhier

Symptoms.—The round-worm is one of the most frequently met with, and is clinically more important than any other human intestinal worm. When it is present in moderate numbers, and occupies its normal position in the small intestines in a person otherwise healthy, there are often no discoverable disorders of structure or function. When present in greater numbers or infesting a delicate person, it is accompanied by the symptoms of irritation of the lining membrane of the alimentary canal, and by consequent impaired nutrition and reflex phenomena. Thus it may be attended with pain in the abdomen, especially in the umbilical region, nausea, impaired or variable appetite, mucous stools, and tumid abdomen. Sometimes, also, pallor of the surface, dilated pupils, swollen eyelids, squinting, irritation of the nostrils, grinding of the teeth during sleep, &c.: indeed, all the allied symptoms which have been attributed to tapeworm. But these are by no means constant effects of the presence of round-worms in the intestine, nor are they peculiar to their irritation. They may be absent when worms are present in considerable numbers; and may be present when no worms infest the patient; or present with the worms but not caused by them. They have, therefore, little or no diagnostic value. Sometimes, however, especially when the intestine contained these worms in very large numbers, they have caused grave local irritation as well as constitutional disturbance, and then post-mortem examination has shown evidences of local superficial congestions and inflammation so closely related to them in extent or position, as to leave no doubt of their causal relation. Thus cases are recorded where numerous round-worms, cohering to each other, gave rise to fatal obstruction and inflammation of the intestines, and others in which they have excited serious and even fatal convulsions in susceptible persons. Although in these latter cases the reflex symptoms are probably in no essential point different from those caused by other irritations, it is important to trace them to the worms, if it can be done, because of the comparative facility with which the exciting cause can be removed. In the rarer cases in which the round-worm wanders during life into distant cavities, organs, or passages, the disorders they induce vary with the parts visited, and may be of great severity, or even terminate fatally.

Diagnosis.—When, for any reason, a patient is suspected to harbor round-worms, it has been a not unfrequent practice to employ the usual treatment for their expulsion—often a rather vigorous one—as a means of diagnosis; and should no worms be passed, it has been assumed that none were present: thus submitting the patient to treatment before the need for it is made out, and assuming, somewhat hastily, that the recognized treatment may be relied upon.

An easy and satisfactory method of diagnosis consists in the microscopic examination of the stools, in which, if the suspected person harbors a mature female, the ova [1] are readily seen. I published a case in the *Medical Times and Gazette* for June 14th, 1856, which so well illustrates the value of this method for diagnosis, and its bearing on treatment, that I venture to give here the following summary of it:—

---

[1] It is curious to note that these ova have been described as cholera corpuscles (*Lancet*, 1849, p. 532) ; and more recently as "choleraphyton," in the *Deutsche Klinik*, 1867.

A girl, aged twelve years, had passed two round-worms before she came under observation, and had complained for six weeks of abdominal pains and disordered digestion. For convenience of observation she was admitted into hospital Feb. 14th, 1855; her stools then contained ova of *Ascaris lumbricoides* (Fig. 23). After nine days, during which she was treated by a mixture of bicarbonate of soda and infusion of quassia, with rest and good diet, she declared herself well, but had passed no worms. For ten days more she was treated by oil of male fern and castor oil, followed by scammony, without effect. For a further period of ten days she took infusions of quassia and senna, also without result. For five weeks more she was given turpentine and castor oil, or turpentine alone, at weekly intervals; and about the third or fourth day after each dose, except the last, she passed one or two worms, generally, but not always, motionless. The ova were still abundant in the stools, but the treatment failing to expel any more worms, she was given *Dolichos pruriens* for four days, until it caused nausea, when it was omitted; but for twelve days more she expelled occasionally one or two worms with the stools. The *Dolichos pruriens* was then repeated for eight days, and again omitted; after which she passed, in the following fortnight, three more worms. The ova were then found to be absent from the stools, and she was discharged. While under treatment she passed in all, seventeen round-worms; but during the last three months and a half she was in perfect health, and would have been discharged but for the observation of the ova in the fæces.

Davaine drew attention to the value of this method of diagnosis in 1857 (*Comptes Rendus Soc. Biologie*, 2ᵉ Série, t. iv. p. 188); and Leuckart says (*Die menschlichen Parisiten*, &c., B. ii. p. 251, 1867), " In the microscopic examination of the fæces we possess a means to determine the presence of the round-worm, which is as easy as it is sure; if it were more generally practised, many errors of diagnosis, and many useless, if not injurious treatments, would be avoided."

The *Etiology* and *Pathology* of the disorders induced by round-worm have appeared on the surface during the previous observations.

Treatment.—The indications are to relieve the irritation of the alimentary canal and improve the general nutrition where that has suffered, but above all things to expel the worms. Many of the substances which have obtained a reputation as anthelmintics have been much used for round-worm, but we have as yet no sufficiently exact knowledge of their action upon the different species of intestinal worms to enable us to estimate their true clinical value in the treatment of *Ascaris lumbricoides*. There exists, however, a very general concurrence of opinion, which I believe to be well founded, in favor of the use of santonica or worm-seed, the unexpanded flower-head of an undetermined species of *Artemisia*, as well as of its active principle, santonin. The dose of worm-seed is from 60 to 120 grains, but it is not much used on account of its inconvenient form; that of santonin which is more used, is from one to three grains twice daily to a child, and from three to six grains for an adult. After a short course of this medicine, an aperient may be given with advantage. It is apt to produce a singular although but temporary perversion of vision if given in too large doses, or for too long a time, objects seeming to be yellow, blue, or green. The urine also may be tinged red after its use. Violent cathartics do not deserve much confidence, nor are the drugs employed for tape-worm (except, perhaps, turpentine) to be trusted to. *Dolichos pruriens* would seem to be worthy of further trial in some cases where santonin is not available, but of the numerous other substances which have been at times recommended for the treatment of *A. lumbricoides*, it is unnecessary to say more here.

The *Prevention* of *Ascaris lumbricoides* cannot be so confidently treated of as was that of *T. solium*, because we are not certain how it enters our bodies; but whether we hold with Leuckart that an intermediate bearer is essential, or with Davaine that it is not, and that we drink the ova in impure water, in all probability the careful cooking of all our foods and drinks would prove a good protection even in those countries and districts in which this pest most abounds. It is not, however, probable that well-filtered water could convey the infection.

ASCARIS MYSPAX (Zeder)

Is the common round-worm of the cat, and is identical with *Ascaris marginata* of the dog (Schneider).

Description.—It is smaller and more slender than *A. lumbricoides*, has two small lateral, cuticular, wing-like appendages near the head. The vulva in the adult female occurs about one-fourth of the whole length from the head. In man it has only been found parasitic in three trustworthy instances, which are recorded by Bellingham, Cobbold, and Leuckart.

OXYURIS VERMICULARIS (Bremser), (Common seat-worm).

Description.—A small whitish fusiform worm, the female attaining $\frac{4}{10}$ in. in length by $\frac{1}{40}$ in. in thickness, and the male about $\frac{1}{6}$ in. in length by $\frac{1}{160}$ in. in thickness (Fig. 24). The head (Fig. 24 *b*, *d*) is furnished with three inconspicuous lips around a terminal mouth, and an elongated vesicular expansion of the cuticular layer on its dorsal and ventral aspects. The œsophagus is continuous, with a muscular stomach containing three teeth, and then follows a simple intestine. The surface is marked by fine transverse rugæ, and the lateral longitudinal bands form a slight angular projection. The female has a long, awl-shaped, caudal extremity (Fig. 24 *c*); the vulva is situated about the junction of the anterior and middle thirds of the body, and conducts to a vagina, a bifid uterus, and this to two tubular ovaries. The male has a blunted tail end furnished with six pairs of papillæ, and a single spiculum communicating with the anal aperture.

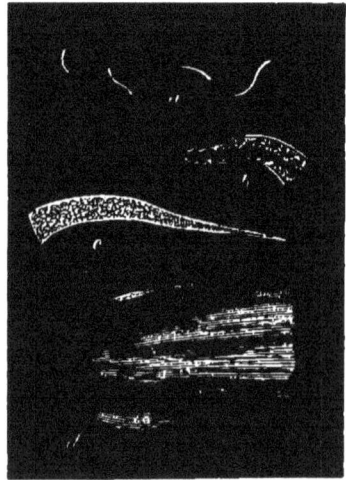

FIG. 24.—*Oxyuris Vermicularis. a* Natural size. *b* Head, magnified. *c* Tail, magnified. *d* Head, more magnified. (Davaine.)

The eggs (Fig. 25) are oval but flattened on one surface, measure $\frac{1}{1700}$ in. by $\frac{1}{4100}$ in., contain at the time of deposition a developing embryo, and have a firm shell consisting of three layers, one of which is absent at one pole, so as to facilitate the escape of the embryo. A moderate estimate allows 10,000 to 12,000 ripe ova for the uterus of a single female.

Life History.—The seat-worm, like the round-worm, is found all over the world, and is perhaps even more frequently met with. It is said to abound particularly in Egypt and in Greenland. It normally inhabits the colon of man only, especially in the neighborhood of the rectum, and is commonly found in large numbers, the males fewer than the females, and it often migrates spontaneously through the anus. The ova are discharged into the intestine of the infested person, and there undergo a further development, so that at the period of their escape with the stools they usually contain a distinctly formed embryo. The frequent spontaneous migrations of the ripe female also often lead to the deposition of the ova upon the skin and hair in the neighborhood of the anus.

The ova deposited with the stools rather rapidly develop under favorable conditions, especially moisture and the warmth of the sun; they are not killed by extreme cold or by desiccation, but a few days' delay in water kills them outright, and under ordinary circumstances they die in a few weeks unless their progress has been arrested by cold or dryness. It does not seem that they hatch in the free state.

Kückenmeister and Vix conceive that all the transformations from the embryo to the adult form take place within the intestine of the infested person without any necessary migration, and at first sight this view seems to receive support from the fact that large numbers of seat-worms are commonly found together, and that various grades of development are there met with. This view, however, is out of accord with the general law of development in parasitic animals, and does not suffice to explain the known facts. Leuckart insists that the emigration of the embryo is a necessary condition of its future development, and has indeed almost proved the correctness of this view by observation and experiment, as well as by powerful arguments. His view is, that ova deposited with the fæces are abundantly and widely scattered in the dry state by winds and other agencies, and then are taken into our stomachs upon uncooked fruits and vegetables and in various other conceivable modes; there exposed to the digestive fluids, the embryos escape, are carried down into the colon and attain the adult form probably in about two weeks. A sort of self infection frequently may take place also; in persons already infested, it is easy to see how the ova upon the skin and hairs near the anus may be conveyed to the mouth by the fingers, after scratching to allay the violent irritation which these small pests produce; and in other modes the eggs may find their way into the stomach from the soiled bed clothes or personal linen. These views explain some long-known facts which are not otherwise so easily understoood; for instance, the great length of time during which some persons suffer from seat-worms, and the liability to relapses notwithstanding repeated treatment; the frequency with which these worms are found inhabiting many members of one family or household, the greater liability of children, of dirty or insane people, and of persons who often eat uncooked fruit and vegetables, as well as the immunity of infants at the breast.

Symptoms.—When only a few seat-worms are present they give rise to no inconvenience, and are usually only accidentally discovered in the stools. When they are more numerous or the patient is more sensitive, they cause an itching or tickling in the anus and its neighborhood, which is sometimes intolerable to the sufferer, especially at a certain hour in the evening. In the female it is peculiarly distressing, from the habit which the worm has of wandering into the vagina; but in both sexes inordinate sexual excitement sometimes is produced. Although there is sometimes evidence of local irritation in the shape of excess of mucus in the fæces

FIG. 25. Ovum of *Oxyuris vermicularis*, from the fæces. (Leuckart.)

and punctiform redness around the anus, the cases of severe convulsion and other nervous disorders which have been referred to the action of seat-worms must be received with much caution.

Diagnosis.—Inspection of the stools will discover the worms; and a microscopic examination will show the ova.

Treatment.—Probably any infected person who adopted the requisite precautions against reinfection from himself or others would get well in a few weeks without treatment by drugs,[1] but this period would be shortened by the use of aperients, and occasionally injections into the rectum of cold water, turpentine and castor-oil with gruel, and of preparations of wormwood, quassia, assafœtida, santonin, &c. Frequent external applications of mercurial or other ointments and lotions likely to kill the embryos might be employed also.

Prevention.—From the foregoing history it may be learned that a sufferer from seat-worms should avoid touching the neighborhood of the anus, should be scrupulously clean [2] in person and clothing; that persons not yet infested should avoid close personal contact, especially in bed, with those who harbor the worms, and should always adopt the caution of eating only well-cooked food.

FAMILY STRONGYLIDES.

DOCHMIUS DUODENALIS (Leuckart).

This minute but dangerous parasite was discovered by Dujardin in 1838 in Northern Italy; its zoological position is scarcely yet settled, but its close affinity to the genus *Dochmius* of Dujardin has been shown by Molin and Leuckart.

FIG. 26.—Male and female *Dochmius duodenalis*, magnified. (Leuckart.)

Description.—It is a small somewhat cylindrical worm: the females measure $\frac{7}{10}$ in. and the males $\frac{4}{10}$ in. in length (Fig. 26). The terminal mouth is surrounded by a dilated capsule directed obliquely backwards and furnished with four large teeth on its longer or ventral border, and with four smaller ones on the opposite or dorsal margin (Fig. 27). The bursa of the male is complex, the spicula two in number. The vulva of the female is placed a little behind the centre. The eggs are oval, measure $\frac{1}{540}$ in. by $\frac{1}{1050}$ in., and when deposited contain a yelk in process of cleavage.

FIG. 27.—Head of *Dochmius duodenalis*, magnified, showing the armature of the mouth capsule. (Leuckart.)

We know as yet but a part of its life history by direct observation, and infer the remainder from that of the better known and very closely allied *D. trigonocephalus* of the dog. The egg, after escaping with the stools, under favorable conditions hatches in a few days, and the embryo enjoys a free existence for a time in mud and

[1] This appears to be a daring statement in the face of past experience, but its probability is measured by the evidence of the life history here given.

[2] The common Hindoo custom of washing after every act of defæcation is worthy of more frequent imitation in this country.

muddy water. It is taken into our stomach by drinking impure water without the intervention of any intermediate bearer, and there it grows and develops to some extent before it passes on into the duodenum or jejunum, where the adult form is assumed. It then attaches itself by its powerfully armed mouth to the villi of the mucous membrane, and sucks the blood of its host. Sometimes, under conditions not yet explained, it becomes encysted between the mucous and muscular coats of the gut. It occurs in warm countries only, has been found in Italy (Dubini), Brazil (Wucherer), and in Egypt (Pruner, Bilharz, Griesinger), where it is a very frequent and dangerous pest, infesting about one-fourth of the entire population. It is present in large numbers together, often by hundreds, sometimes by thousands, and then may cause frequent and dangerous hæmorrhages into the bowels, followed by an anæmic condition, which is often fatal, and to which the name of Egyptian chlorosis had been given before Griesinger pointed out its true nature.

Doubtless its ova might be found in the stools of infested persons, but of the treatment which should follow a diagnosis so established little can be said, except that Griesinger recommends turpentine, and that santonin and such other substances as are believed to expel nemetode worms should be tried. Care should also be taken to consume only pure water or drinks which have been boiled, so as to avoid reinfection, and the patient might then be fairly expected to outlive the worm.

Although to the practitioner in Britain this parasite is not of practical import, it seems so probable that it may be found in India or some of the tropical British colonies, that I have ventured to include it here.

FAMILY TRICHOTRACHELIDES.

TRICHOCEPHALUS DISPAR (Rudolphi).

Description.—The female measures about $\frac{4}{8}$ in. the male about $\frac{4}{8}$ in. in length. The anterior three-fifths of the body are threadlike, measuring $\frac{1}{1600}$ in. only in thickness, and bear a simple terminal mouth without papillæ. The posterior two-fifths, about $\frac{1}{84}$ in. in thickness, contain the generative organs and the intestinal canal; in the male, it is spirally coiled, in the female slightly curved (Fig. 28). The caudal extremity is rounded off in the male, and bears a single blunt spiculum in a tubular protrusile sheath which is furnished with teeth. The vulva in the female opens about the level of the stomach into a vagina, the walls of which are furnished with teeth, and often prolapse. The large uterus contains thousands of eggs, which are elliptical in form, and have a nipple-shaped projection at each end. They measure $\frac{1}{450}$ in. by $\frac{1}{1120}$ in. (Fig. 29), and have a firm brownish yellow shell, wanting at each pole, so as to leave an aperture which is closed by a firm transparent nipple-shaped plug. As found in the stools the yelk shows no trace of commencing development.

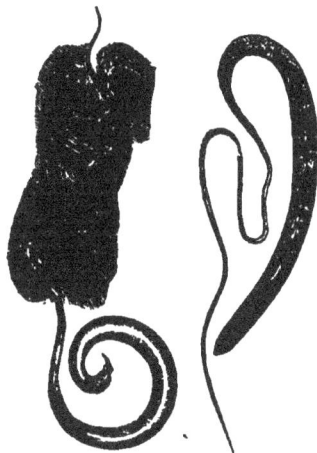

FIG. 28.—Male and female *Trichocephalus dispar*, magnified. (Leuckart.)

Life History.—The *Trichocephalus crenatus* of the pig, and also that found in some monkeys, is probably the same as our *T. dispar.* It is met with in most, if not in all, European countries; in Syria, Egypt, and North America; it abounds in Italy, and in some Eastern lands; but is said to be comparatively rare in Copenhagen and in London (Cobbold). It does not generally occur in large numbers together, although sometimes hundreds have been found. The head of the colon is its chosen residence, but occasionally it is met with in the intestines near. During

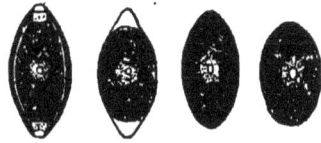

FIG. 29.—Ovum of *T. dispar.*

the life of its host, it attaches itself by thrusting its long whip-like neck into the mucous membrane. The ova deposited with the stools, like those of *Ascaris lumbricoides*, very slowly develop normally in damp earth or water, so that in warm weather and under favorable conditions the embryo is formed in about four or five months; but in cold weather or exposed to temporary drought it requires a year and a half or more. In this state the embryo remains, and neither develops further nor leaves the shell to become free. (Davaine has preserved them alive in this state for four years.) From Leuckart's experiments upon the *Trichocephali* of sheep and pigs, it is highly probable that no intermediate bearer intervenes, but that we swallow the ova with their contained embryos in some accidental manner, as dust upon uncooked fruit, vegetables, &c. &c., and that the embryos escape into our stomachs after partial digestion of the shells, develop somewhat, and then travel onwards to the colon, where they become sexually mature in four or five weeks.

No symptoms are known to be caused by *T. dispar,* although some writers have attributed severe reflex disorders to them when present in large numbers.[1] The worm may be readily shown to be present by finding the ova in the stools. A satisfactory treatment by drugs is not yet known, but there is consolation in the reflection that the parasite has probably a short duration of life, and that we may prevent further infection by avoiding uncooked foods and drinking pure water.

---

[1] When Röderer and Wagler, about a century ago rediscovered this worm, Morgagni's prior observation having been forgotten, they supposed that it produced the typhoid fever then prevailing at Göttingen. It is not difficult to see how such an error arose, the worms having been found in the bodies of most of the victims of the fever, and nearly coinciding in seat with the local manifestations of the disease. In connection with this, it is noteworthy that the more modern theory of the etiology of typhoid fever receives an indirect support from the fact that every person who is shown to be infested with those very common Entozoa *Oxyuris vermicularis* or *Trichocephalus dispar* is thereby demonstrated to have swallowed minute portions of his own or another person's fæces.

PERITONITIS.

By John Richard Wardell, M.D., F.R.C.P.

Definition.—An inflammation of the serous membrane which invests the abdominal organs and lines the abdominal cavity. It may be partial or limited, or it may be diffused over the entire inner surface of the peritoneal sac. Effusion is almost the invariable consequence, and examination after death discovers serum, albuminous exudation, sero-purulent, purulent, or sero-sanguineous fluid and organized adhesions.

Preliminary Observations.—Inflammation of the peritoneum is characterized by the kind of phenomena which are exemplified in the inflammation of the other serous membranes. It may occur at all ages, in every description of temperament, and under the most varied conditions of the system. It attacks the earliest infancy as well as the adult and those in advanced life, and both sexes are equally liable to the affection. It happens to the robust and plethoric, the cachectic and attenuated, and also to those whose constitution has been undermined and broken down; and whenever its distinguishing symptoms are really proclaimed it is one of the most formidable maladies with which the physician has to deal. It may come on suddenly with apparent and easily recognized symptoms, or it may supervene slowly and insidiously, and continue for a time without being detected. It may be primary when it is difficult or absolutely incapable of connection with any foregoing or coëtaneous disease. It may be consecutive upon, or symptomatic of, some other morbid condition. It may present the sthenic or asthenic form. And it may be met with only in sporadic cases, or prevail as an epidemic. Every example of the complaint will, if carefully studied, exhibit some peculiarity—some cognizable difference in its physiognomy, if such term may be employed—dependent upon the degree of mal-nutrition, or the metamorphosis of the tissues, upon the operation of external agencies, the time of life, the amount of vital power, and the idiosyncrasies of the patient. It will be modified by the state of the depurative organs, and especially by that of the kidneys, because those deleterious and effete matters which ought to be carried off by the renal organs, when retained in the circulation, are particularly prone to institute the inflammatory process in serous membranes. When the disease is regarded in all its phases and its cardinal signs are duly observed, it exhibits a train of phenomena peculiar to its own morbid action; and if Peritonitis, like pneumonia and certain other diseases, which formerly had always accorded to them an essentiality, is not to be deemed an essential complaint—a nosological entity, as some modern pathologists maintain—it certainly from its importance demands a distinct place in a comprehensive work like that of "The System of Medicine."

The other authors did not distinguish the inflammation of this mem-

12

brane as apart and disconnected, but only as associated and confounded with the inflamed condition of subjacent organs and tissues, nor was it until the close of the last century that this distinction was made. Since that time the researches of Broussais, Bichat, Barron, Hodgkin, and more recently of Habershon, have extended our information, and given much precision to our knowledge on the subject. Sauvages remarks:—*Enteritis mesenterica* (Peritonitis) *difficillime distinguitur ab enteritide, quacum etiam sæpe complicatur.*"[1]   Cullen says it is difficult to say by what symptoms it can be recognized, and more recent authors have expressed themselves in similar language; but, as will hereafter be shown, it unquestionably displays features by which it can be diagnosticated. John Hunter thus delivers himself on this subject:—" If the peritoneum which lines the cavity of the abdomen inflames, its inflammation does not affect the parietes of the abdomen; or if the peritoneum covering any of the viscera is inflamed, it does not affect the viscera. Thus the peritoneum shall be universally inflamed, as in puerperal fever, yet the parietes of the abdomen and the proper coats of the intestines shall not be affected."[2]   That these propositions are sometimes verfied it cannot be denied, but according to my own experience in Peritonitis which has existed for a time, it well-nigh always happens that some of the organs and structures which it covers reveal the inflammatory products. Dr. Habershon, in a valuable article[3] on the etiology and treatment of Peritonitis, speaks with much boldness and decision on this question, and he bases his conclusions on the trustworthy grounds of accumulated facts.   " In 3,752 inspections recorded at Guy's Hospital," says this physician, " during twenty-five years 500 instances of Peritonitis occur, but we cannot find a single case thoroughly detailed where the disease could be correctly regarded as existing solely in the peritoneal serous membrane." He then divides them, firstly, into Peritonitis by extension from diseased viscera or direct injury; secondly, into those connected with blood changes, as in albuminuria, pyæmia, puerperal fever, and erysipelas; and thirdly, into those caused by nutritive change, as in struma and cancer. This author then contemplates the affection, so-called Peritonitis, as nothing more than the local evidence of antecedent morbid changes pervading the whole system. Dr. Sieveking says it is the climax of nutritive derangements, certainly not to be sought for primarily, in the serous investment of the intestines.[4]   The former of these authorities denies that it is *ever* idiopathic, but he would almost seem to discard that term from pathological phraseology, as he conceives it can hardly with correctness be applied to any disease spontaneously instituted within the organism, and not dependent upon external noxious agencies or parasites.

Sometimes the lesion is but partial, in other instances it extends over the entire membrane, and doubtless it is at the outset only that it is limited, and that its diffusion gradually supervenes. Its closest analogies are pleuritis and pericarditis, and like these affections it is broadly distinguished by its tendency to effusion, adhesions by coagulable lymph, or the deposition of purulent or sero-purulent fluid. The pathologic conditions consequent upon Peritonitis, as of the other maladies now instanced, are sometimes inceptive of further disease, or they may be de-

---

[1] Classis iii. gen. xv. sp. iv.
[2] On the Blood. p. 244.
[3] Medico-Chirurgical Review, No. xliii.
[4] Croonian Lectures, *British Medical Journal*, April 14, 1866.

fensive against worst results; [1] they may eventuate in the union or binding down of organs and parts whereby their functions are seriously or even fatally interfered with; or this same tendency to albuminous exudation may, as in some instances of perforation, be conservative of life, the plastic deposit being the means whereby nature essays to effect reparation.[1] But these and kindred considerations will be more fully considered when I speak of the pathology and morbid anatomy of the disease.

ETIOLOGY.—The causes of Peritonitis are often traceable to wet and cold, damp feet, damp beds, chill winds, sudden alternations of temperature, as when, after being in a heated atmosphere, the body is rapidly cooled, or to excessive fatigue—in fact to such general influences as are concerned in the production of inflammation in other viscera. It may, in a more direct manner, be induced—in a mechanical way—by invagination, strangulated hernia, surgical operations (as in paracentesis abdominis, and ovariotomy); by contusions, bruises, the wounds of cutting or blunt instruments; by displacement of some of the internal organs, or some unusual stretching or laceration of the membrane;—by the extrusion of certain matters into the serous sac, as in hepatic or splenic abscess, rupture of the stomach, bile-ducts, spleen, uterus, urinary bladder, ureters, the ovary or some part of the sub-diaphragmatic digestive tube. It may follow or be associated with the acute disease of some organ by contiguity of structure, as in gastritis, hepatitis, splenitis, in dysentery, or in typhoid fever when the lower third of the ileum or the vermiform appendix is ulcerated. Sometimes tumors, extra-uterine conceptions, or malignant growths by the induction of pressure, or ulcerative absorption, give rise to it. The abrupt suppression of habitual discharges, more especially of the catamenia and lochia, and the sudden retrocession of cutaneous eruptions, have been enumerated; and contamination of the blood itself, resulting from the altered and imperfect action of certain of the excreting organs, enters, there are good grounds for believing, far more frequently and far more importantly as an element in the causation than has hitherto been supposed. Indeed, many attacks which we regard as idiopathic are dependent upon a common cause in the organism, but this membrane may sometimes have a greater proclivity to the condition of inflammation than any other part. Sometimes Peritonitis is metastatic of rheumatism, erysipelas, and the exanthematous fevers. Broussais repeatedly knew it succeed intermittent fever, and it is occasionally connected with fevers of a malignant type.

SYMPTOMATOLOGY.—The invasion is often sudden, but the attack may come on slowly and covertly. In the acute sthenic form there are generally rigors, followed by heat and flushings, a feeling of lassitude, aching of the limbs, head, or back, a sense of constriction and uneasiness at the epigastrium, thirst, nausea, and acute pains at some, especially the lower, part of the belly. Pressure on the abdomen, coughing, sneezing, the evacuation of the bladder or bowels, or even the erect position, augments the pain; indeed whatever produces weight upon or stretches the membrane, of necessity aggravates the suffering. The pain is at first localized, but it soon becomes diffused over the entire abdomen, and is a prominent sign. As the disease progresses, the pulse becomes quick, hard, sharp, and tense, and rises from 120 to 130 in the minute. In some exceptional cases it does not ascend to more than 80 or 90, and is of tolerably full volume; but as the rule it is firm, small, and cordy. The

---

[1] Sir Thomas Watson.

pulse is not always, however, a sure guide, as most serious attacks may be progressing under all conditions of the arterial circulation; and even pain on pressure, the most trustworthy of all individual symptoms, is not invariably to be relied upon, because it is not uniformly commensurate with the amount of lesion which really obtains. The tongue is mostly moist and covered with a whitish creamy mucus, but occasionally it is dry. The bowels have a tendency to be confined, and the urine is scanty and high-colored. The skin is hot and dry at the earlier period of the disease, and becomes cool and bedewed with a clammy sweat before dissolution. The patient lies in the supine posture with knees drawn up, and cannot turn on either side without increase of pain. He will say that he experiences a feeling of heat, pricking, cutting, or soreness in his inside; involuntarily he relaxes the abdominal muscles, and sometimes fomentations, and even the weight of the bed-clothes cannot be borne. The breathing becomes quick, shallow, and almost entirely thoracic, and instead of being 18 or 20 it may be 50 or even 60 in the minute. The downward pressure of the diaphragm is instinctively as much as possible avoided, because it moves the abdominal organs, and all movement gives pain. The passage of flatus along the bowels is followed by the same effect. With regard to the pain, which is a cardinal sign, it presents some differences; sometimes it is permanent, in other cases it is paroxysmal, assuming a spasmodic character, and in a few rare examples it is not present in marked degree. As the rule, it is the chief and most reliable symptom.

There is always between this disease and the features great sympathy. The face becomes pale, the cheeks collapse, and the eyes seem set and sunken in their foramina. It assumes the Facies Hippocratica, or what the French term the *Facies Grippée*. Nausea and vomiting often come on with the other symptoms, the ejected matters being a mucoid, biliary fluid; or, in the case of obstructed bowels, the vomited matters may be stercoraceous. Tympanitis is never absent, and often very distressing. The loss of tone in the muscular coat, and the irritation which is conferred on the mucous surface of the alimentary canal, account for such condition. The distention varies in degree. In those whose bodies are flabby and resistless it is often excessive, whilst in the robust and muscular it is in less amount. If the diaphragmatic covering becomes inflamed, singultus often occurs; when the serous coat of the stomach is involved, sickness is urgent; if that of the urinary bladder, there is strangury; and the inflamed tunic of the kidneys will produce ischuria renalis. Percussion elicits the loud tympanitic note, especially in the umbilical and epigastric regions. When there is effusion of serum—which, of course, gravitates to the lower parts — the line of dulness can be most distinctly observed, and it is in some measure altered according to the position of the body. Palpation can only be had recourse to with great care, as the extreme tenderness and muscular resistance prevent much manual examination. When effusion has taken place, and coagulable lymph has matted the intestines together and formed roughened deposits on the liver, spleen, or some tumor, and when albuminous concretions adhere to the parietal peritoneum, the flat hand laid on the abdomen feels a peculiar thrill or vibration, which is most distinct during inspiration. This sign only obtains when the lymph is thrown out on a resisting basis. Auscultation discovers a creaking friction sound, which is variable in character and intensity, and can only be present for a short time, as of course, on the advent of adhesion, nothing can be heard. The physical signs of pericarditis and pleuritis are far more common. Death is ushered in by quick and thready pulse, cold and clammy surface, loss of heat in the feet and legs, accelerated and labored breathing and general declension of power, the mind being often clear and collected to the last. Pemberton says the patient frequently expires on the sixth, seventh, or eighth day. But it is equally true that the fatal issue often occurs in two or three days. In puerperal Peritonitis the average duration of the disease has been shown to be about thirty hours, and sometimes, as in perforation, it may be even less than ten hours. When the affection assumes a more chronic form, the patient may live so long as thirty or forty days.

The asthenic type of Peritonitis occurs in the cachectic, and those whose vital powers have been undermined by some previous disease. It is that form which is seen as metastatic of erysipelas and rheumatism, and in connection with the exanthems, malignant fevers, puerperal women, and when there is perforation of some part of the digestive tube. It proclaims contamination of the blood and want of vital power. The effusion is sudden, large in quantity, of debased character, and notably deficient in organizable plasma. The pulse is soft and feeble, the surface soon becomes moist, and all the phenomena proclaim declension of vitality.

When the disease terminates by *resolution*, a gradual improvement of all the symptoms becomes observable. The symptomatic fever declines, the pain is less urgent, and pressure can be borne on the abdomen; the skin is moderately moist, but not below the ordinary temperature; the tongue looks cleaner; the pulse is slower, fuller, and softer; the respiration is more normal, being less frequent and not so thoracic; the alvine evacuations are freer and more natural; and there is generally a copious secretion of urine, which contains an abundance of lateritious deposits. Sometimes moderate diarrhœa or diaphoresis are critical discharges. The sickness and vomiting cease, the tympanitis and feeling of distention obtain in less degree, and the patient can extend his legs and lie on either side with more freedom and ease. Lastly, the countenance, which had hitherto been so faithful an index of the complaint, looks calmer and more natural, it having lost much of the sunken, collapsed appearance above described.

One of the most frequent results is *effusion;* indeed, the affection cannot assume a well-marked and typical character without one or other of the inflammatory products being thrown out, and these, as to their proportion and quality, are varied in every individual example. In the earlier stage of the attack the effusion is but small, and not such as in marked manner to increase the size of the abdomen. It gravitates into the pelvis and the iliac fossæ. It can be detected by percussion over the lower parts of the belly, and there are general signs which indicate its presence. When it increases, the pain becomes a less prominent symptom, the pulse is softer, there is a feeling of weight and dragging in the body, chilliness and a diminution of animal heat, the extremities having a tendency to become cool. In such cases as are metastatic of some other complaint, the effusion is much more rapidly generated and the serous proportion is relatively very large. Andral records an illustration which was metastatic of rheumatism, and which ran through its course to a fatal termination in three days, and the autopsy showed an enormous quantity of serum tinted with the coloring matter of the blood, and some floating flocculi and false membranes. When pus is secreted, rigors are a common symptom, with febrile exacerbation in the evening, and the pulse is quicker.

It is not, if in any notable quantity, absorbed. It finds an exit either by forming an ulcerated opening into the bowel, which is always fatal, or, which is much more common, it establishes a fistulous passage by way of the psoas muscle, or through some part of the abdominal walls. In this tendency to appear at the surface it seems to obey the law of an ordinary abscess.

Inflammation of the peritoneum rarely ends in *gangrene*, and it is still more rare for any considerable portion of the membrane to become gangrenous. When it has come on, it has generally been at or about the vermiform appendix, or when some part of the bowel has been unduly stretched or strangulated; and, according to Abercrombie, it is invariably accompanied with false membranes. The sudden cessation of pain, singultus, coldness of the surface, thready compressible pulse, general declension of strength, and the Hippocratic countenance, are indicative of this condition.

Sometimes the acute gradually passes into the chronic form, when, as before remarked, the patient does not die until after five or six weeks. He may live even several months. In such cases the effusion may not be absorbed nor yet evacuated, or a fistulous communication may have been produced, and all the conditions of asthenia usher in the mortal event. Again, in other examples, the serous fluid will be absorbed, the adhesions become firm and organized, or the sero-purulent or purulent matter be discharged, and slow recovery result.

The phases which the inflammation of this membrane may assume are very varied; and it is only by the study of a large number of examples that the physician can anticipate and comprehend the modes of its progress. Sometimes that cardinal symptom, pain, upon which such emphasis has been laid, only obtains at the outset; and notwithstanding its subsidence, the malady goes on. Occasionally, as in pleuritis, there may be little or no pain from first to last, whilst rigors and hectic and wasting pronounce still the seriousness of the case at a time long after that period when danger is generally thought to have passed away, and a large collection of pus is contained in the cavity; or the acute symptoms may rapidly subside under a properly directed antiphlogistic treatment, and the condition of simple ascites will only appear to be present; again, disease instituted in some of the abdominal organs will greatly modify the affection after it has become chronic. In this state adhesions alter the configuration of the abdomen by large masses of fibrin being deposited together, by the soldering of the intestinal convolutions, the agglomeration of one organ to another, or by the formation of separate collections of matter in distinct septa resembling independent abscesses. It sometimes happens, too, that the belly becomes soft and flabby, and, instead of improvement succeeding this disappearance of the tension, convalescence is slow and protracted. From what has now been said, it is obvious that the chronic condition is far from being uniform in its phenomena, and that the pathological changes may be diverse and multiform.

VARIETIES.—Broussais and some other authorities speak of the induction of Peritonitis by the exudation of blood into the abdominal cavity without solution of continuity in any of the blood-vessels. I have never seen such an instance, and these examples must be extremely rare. Such sparse exceptions are to be associated with hæmorrhagic diathesis, the predisposing causes being the sanguine temperament and a marked tendency to inflammatory complaints. According to Broussais, the pulse is at first full, but soon becomes soft and compressible, the pain very acute, often intermittent, and coldness of the extremities and convulsions quickly

close the scene.[1] Laennec was one of the first to draw attention to hæmorrhagic exudations of serous membranes, and Rokitansky attributes such
tendency to the tubercular cachexia, the diseased condition of the blood
resulting from cirrhosis of the liver, the scorbutic constitution, and the
dyscrasia of drunkards. The effect of specific poisons, such as induce the
various febrile diseases, and that anomalous condition of the blood now
spoken of in which its fibrinous constituent is diminished, and its serous
part augmented, are to be enumerated in the causation of this hæmorrhagic
exudation. When the blood having this origin is discovered in the peritoneal sac, it is in large quantity, very red, and in varying proportions
mixed with serum.

There is another description of Peritonitis which systematic writers
have recorded, and to which the name of *latency* has been given. It has
been said to attack those laboring under some other ailment, the feeble
and attenuated, the aged, the insane, and such as exhibit a low degree of
vitality. Its symptoms at the outset are masked and difficult of recognition, and, when recognized of the asthenic type, the features present those
distinguishing traits before insisted upon as being characteristic of this
complaint. It is evident that such examples are nothing more nor less
than secondary affections like unto pneumonia in albuminuria, pleuro-
pneumonia when intercurrent in phthisis, pericarditis in rheumatism, and
arachnitis in continued fever.

Non-plastic or Erysipelatous Peritonitis.—This is seen as the sequel
or complication of the exanthems, in adynamic fevers, and in puerperal
Peritonitis. Its essential condition is some hæmic change, and it is
characterized by asthenia. It is met with in worn-out and undermined
constitutions, in the unhealthy, and in those who have had some other
malady. Its supervention is sudden, and it runs its course with great
celerity. It does not bear an antiphlogistic or lowering treatment, and
is only benefited by stimulating and sustaining remedies. According to
Abercrombie, "the symptoms are sometimes slight and insidious, but
sometimes very severe; and they are chiefly distinguished by the rapidity
with which they run their course, and by a remarkable sinking of the vital
powers which occurs from an early period, and often prevents the adoption
of any active treatment. A remarkable circumstance in the history of
this affection is its connection with erysipelas, or with other diseases of
an erysipelatous character."[2] Illustrative of this form of the complaint
he gives the instance of a woman who had erysipelatous inflammation of
the throat, who was very suddenly seized with abdominal pain and vomiting, and who gradually sank in forty-eight hours. The necroscopy discovered a large quantity of pus in the peritoneal sac. And he gives other
and similar examples. This physician also refers to an epidemic of erysipelatous character which occurred amongst the children in the Merchants'
Hospital, Edinburgh, in 1824. The disease was of mild type. In all the
cases there was throat affection, consisting of a raw, red appearance,
swelling, and aphthous crusts. Two of the little patients speedily sank,
and inspection revealed pus in the abdominal cavity. Abercrombie draws
a comparison between this epidemic and one of diphtheritic, as it was then
named, which appeared two years afterwards, and he believed them to be
congeners. The correctness of this opinion later years have confirmed.
Between diphtheria and erysipelas there is great resemblance. They are

---

[1] Broussais, Histoire des Phlegmasies ou Inflammations chroniques.
[2] Pathological Researches on the Diseases of the Abdominal Viscera, 3d edit. p. 181.

both referrible to general blood change, and, as it has been well remarked, are associated with a large group of maladies which stand in close relation with pyæmia.[1] The kind of Peritonitis spoken of occurs with a depressed vitalism, consequent upon toxæmic agents imbibed from without or formed within the organism by its own power of genesis; and the term non-plastic well applies to the ostensible difference which there is between this type, deficient in organizable plasma, and the adhesive form of inflammation.

Perforation of the Peritoneal Membrane.—There is no form of Peritonitis which is so fearful and fatal as that in which there has been positive solution of continuity of the membrane, because this accident generally implies the extrusion of some secretion or fluid or substance into the serous cavity. Several of the older authors mention this occurrence, and some vaguely attribute such openings to worms—a *possibility*, as we know from Andral's case, in which lumbrici passed into the cavity; but this event is exceedingly rare. There is no doubt that in nearly all these recorded instances the real cause of such perforations was ulcerative destruction, or cadaveric change, which former writers had not recognized with that facility and certitude which distinguish the acquisition of modern pathologists. Perforation may be produced in a great variety of ways, by penetrating wounds made by sharp or blunt instruments, the crushing effect of accidents, lacerating the solid or hollow viscera, or the parietal peritoneum; corrosive poisons, the giving way of the uterine walls during parturition, the softening of a fibrous tumor attached to the uterus and the contents being extravasated; the bursting of a Graafian vesicle, of a mesenteric gland, of a tubercular deposit, of the urinary or gall-bladder; from calculi, from the evacuation of some collection of purulent matter, as in empyema; burrowing through the diaphragm, in abscess, as before remarked, of the liver, spleen, or kidney, in pelvic abscess, and from other causes. Mr. Hulke lately recorded an instance of renal abscess bursting into the peritoneal sac, which occurred in an unhealthy-looking maidservant who was admitted into the Middlesex Hospital for hip disease, and which ended fatally. The inspection discovered puriform serum in the peritoneal cavity, and the peritoneal surfaces were coated with a soft yellow lymph. The right kidney was a mere sacculated pouch, and it was ruptured at its upper end.[2] The more common cause of perforation is ulceration, commencing in the mucous membrane, of some portion of the digestive tube, and penetrating through the muscular and serous coats. It may be referrible to softening of the intestinal wall (*ramollissement gélatiniforme*), or to cancerous disease, especially when the cancerous deposit encroaches upon, or absolutely blocks up, the passage. When the accident is from this cause, it is mostly observed in the stomach, colon, or cæcum.

The symptoms are sudden, often violent. Frequently the patient at once falls into collapse. Andral says, that sudden increase of prostration and rapid change of the features are sometimes the only symptoms denoting the accident of perforation. Sometimes there is febrile excitement, as evinced by increased heat of surface, hard pulse, and urgent thirst. In the great majority of cases remedies seem inoperative; the disease rapidly becomes diffused over the surface of the sac; whilst vomiting, dorsal decubitus, quick and feeble pulse, loss of animal heat, and sunken and collapsed features, too truly indicate the powerful impress

[1] Dr. Russell Reynolds, art. Erysipelas, vol. i.
[2] *Lancet*, Jan. 23, 1866.

which has been made upon the circulatory and nervous systems, the mental faculties, generally, remaining unaffected to the last.   In those very exceptional cases in which recovery does take place the vomiting begins to subside, the distention to decline; the pulse becomes softer, fuller and slower; the face is less haggard, the patient sleeps more tranquilly, and the temperature of the body is more natural.

When the stomach is the seat of perforation, as it sometimes is, by simple or specific ulcer, the phenomena are precisely those which obtain when any other part of the sub-diaphragmatic tube gives way.   Ulceration of this organ is most frequent in females.   Dr. Brinton found that in 654 cases 440 were in females, and 214 in males.   He also says that in the former sex one-half occurred between the ages of 14 and 20.[1]   It happens to children.   Dr. Lee knew perforation of the stomach of a girl of eight, and in that of a boy of nine years of age.   The opening is most frequent at the splenic end, and that part is also most prone to gelatiniform softening. It may give rise to hæmorrhage.   Habershon gives an example in which the splenic and pancreatic arteries were opened.   It does not absolutely follow that death shall always eventuate, because adhesion may take place between the point of ulceration and the abdominal walls, or one of the solid viscera, or a communication may be established between the stomach and the colon, or the duodenum, or a gastric fistula may be formed externally, or through the diaphragm into the thorax.   The last two named are very uncommon, but possible contingencies.   Abercrombie gives an example of the kind of Peritonitis now considered.   A young woman had been affected with dyspeptic symptoms and epigastric pain for some months.   On Nov. 26th, 1824, she was heard to scream violently, and when approached was unable to express her feelings except by violently pressing her hand against the pit of the stomach.   The abdomen became tender and distended, and she continued in extreme suffering till the 27th, when she died twenty-nine hours after the attack.   On the inspection of the body the cavity of the peritoneum was distended with air, and likewise contained upwards of eight pounds of fluid of whitish color and fœtid smell.   There was slight but extensive inflammatory deposition on the surface of the intestines, producing adhesion to each other, and to the parietes of the abdomen.   In the small curvature of the stomach was a perforation which admitted the point of the little finger.[2]   This author gives another case in the person of an elderly gentleman, who was suddenly seized with excruciating pain at the stomach, accompanied by vomiting, coldness, and quick pulse.   The abdomen became tense and tender, and he sank in thirty hours.   Necroscopy exhibited near to the pyloric opening an ulcerated hole larger than a shilling, to which the liver formed a base, and a little below the perforation of the calibre of a quill through which the contents of the stomach had escaped and caused fatal Peritonitis.

---

[1] Dr. Brinton gives the following relative proportions per cent. of the locality of perforations which ended fatally by Peritonitis :—

Posterior Surface	2
Pyloric Sac	10
Middle	13
Lesser Curvature	18
Anterior and Posterior Surface at once	28
Cardiac Extremity	40
Anterior Surface	85

[2] Abercrombie's Diseases of Stomach, 3d edit. p. 34.

The duodenum is less liable to this accident than the stomach; but its serous tunic does sometimes give way under the ulcerative process.   Mr. Curling was the first to observe that the glands of Brunner are apt to pass into ulceration during the progress of severe burns, and from this cause Peritonitis may in a secondary manner result.   In twenty-two autopsies made by Louis in enteric fever, in only two cases was the villous surface of the duodenum found ulcerated.   In fifteen examples of that disease examined by Jenner, and in twenty by Murchison, no morbid condition was detected in this organ.   Its ulceration in all its characteristics and consequences very nearly resembled that described of the stomach.   Habershon says several cases have come under his observation, the early symptoms of the ulceration being slight until fatal Peritonitis had been set up by perforation.   In other instances violent vomiting produced the accident. Hodgkin relates the instance of a young woman, aged twenty-four, who was admitted into Guy's with urgent vomiting, small and feeble pulse, and who shortly after died of fatal Peritonitis caused by a small ulcer in the duodenum.   Habershon gives an interesting example in a young woman, aged eighteen, admitted into Guy's February 19th, and who died October 4th, 1860.   At first the prominent symptom was vomiting; after a time diarrhœa came on, and the emaciation increased.   Examination of the body showed behind the first portion of the duodenum and close to the pancreas a collection of offensive pus, and a perforation a quarter of an inch in diameter was discovered.   From the histories of six cases recorded by Dr. Andrew Clark,[1] he concludes that the event is sudden, after food, and that the pain never leaves its place of origin.   In the examples given by this physician there was no sensation of something having given way, nor of heat diffusing itself over the belly.   This organ is more frequently perforated by secondary than primary disease.   The malignancy of neighboring viscera is sometimes extended to its parietes, as in cancer of the stomach, liver, spleen, pancreas and lymphatic glands, and its consequent rupture is followed by Peritonitis, which ends fatally.

With regard to the jejunum it is rarely found morbid, and assuredly no part of the digestive tube possesses such an immunity from disease. I have known no instance of its perforation.   Neumann and Hufeland, however, have recorded an example of this event.

Perforation more frequently occurs in the lower third of the ileum, and near to the ileo-cæcal valve, than in any other part of the intestines.   Of ten cases by Louis, it was within a foot of the valve.   Of ten cases given by Stokes, in nine it was within twelve inches of the valve, and one was in the cæcum.   Of eleven by Murchison, nine were within twelve, and two within eighteen inches of the same place.   Bartlett saw it forty-four, and Bristowe seventy-two inches from the same place.   The parts next in order of prevalence are the cæcum and vermiform appendix.   Louis was one of the earliest observers of the facts now noticed.   It has long been broadly and familiarly known that the agminate glands which are proper to the ileum, and the solitary glands which are scattered throughout the villous coat of the digestive tube, are in enteric fever very prone to take on the ulcerative condition, more especially the patches of Peyer, and occasionally it happens that after the mucous and muscular coats have been destroyed, the peritoneum gives way.   These glands are not in like manner predisposed to disease in the course of any other acute affection. The vermiform appendix has in repetition been found the seat of fatal

---

[1] *British Medical Journal*, June 22, 1867.

Peritonitis, not only in enteric fever, when sometimes only a very minute orifice can be discovered, but from the impaction of some foreign body, as the seed of fruit, a kernel, a piece of bone, a piece of indurated fæcal matter, or even the single bristle of a tooth-brush. Of eight cases of perforation given by Louis, seven were in the young and vigorous, and it may here be observed that more recent writers, as Jenner, Murchison, and Bristowe, have shown that it chiefly occurs between the ages of fifteen and twenty. Of the eight cases by Louis, with a single exception, the disease commenced with continued fever, nor did the febrile phenomena assume any severity of character until the advent of the perforation. In four there had been diarrhœa, but only in one were the bowels much harassed. Tweedie says the state of the bowels, either as to the presence or absence of diarrhœa, is not to be depended upon, as it sometimes happens that the evacuations are healthy when the bowel gives way. Three were quite convalescent when the opening occurred, and a fourth appeared to have fully recovered from an attack of enteritis.

Since Louis wrote his account, much information has been accumulated on this particular subject. It is now well known to all who have made the various forms of fever a special study, that there is no precise correlation between the gravity of febrile symptoms and the occurrence of perforation. The diarrhœa may have been a distressing and persistent symptom, and yet the points of ulceration may not have been either numerous or deep; on the other hand, in cases regarded as mild forms of fever the bowel may very unexpectedly burst, and this event is generally at a later date of the attack, or during convalescence. Tweedie has known it take place when the patient has so far recovered as to leave the house. Dr. Murchison lately published an apt illustration.[1] Some time ago I had under my care a girl in enteric fever who became quite convalescent, and at the end of six weeks, after eating a hearty meal of solid food, Peritonitis supervened, and she died in twenty-two hours. Peacock saw it come on so soon as the eighth, and Murchison on the ninth day of fever. Louis noticed it so late as the forty-second, and Jenner on the forty-sixth day. Of thirty-two cases given by Murchison, perforation occurred during the second week in eight cases; during the third week in six, during the fourth week in nine, and after the fourth week in nine.[2] Louis says, if in acute disease, and in an unexpected manner, a violent pain in the abdomen supervenes; if this pain is exasperated by pressure accompanied by rapid alteration of the features, and more or less promptly followed by nausea and vomiting, we may believe and announce that there is perforation of the intestine.[3] Pain is not a symptom in all cases continuous up to death. It sometimes notably abates, and in exceptional examples ceases entirely for several hours before dissolution. Jenner saw a patient in whom there was no pain at all, vomiting and cold extremities being the only symptoms. Tweedie asserts that the symptoms of this event are not uniformly well pronounced. The accident may be masked by delirium so considerably that the time of perforation and its absolute occurrence may be uncertain.

Dr. Stokes gives particulars relative to nine cases which occurred under his own observation.[4] These happened during fever; one in catarrhal fever, two after acute enteritis, and in one case hypercatharsis produced by an overdose of salts was the cause. In several of these nine instances

[1] *British Medical Journal,* Dec. 2, 1865.   [2] On Fever, p. 508.
[3] *Recherches Anatomice-Pathologiques.*   [4] Cyclop. Pract. Med.

there had been diarrhœa. He also comments upon a fact worthy of notice, that in three were produced irritation of the bladder and inability
to pass urine. In all, inspection revealed ulceration of the muciparous
glands; and respecting the time which the patient lived after the initiatory symptoms of perforation, it varied from twelve to one hundred and
twenty hours. Stokes also says that the average duration, deduced from
nineteen cases which he had collected from various sources, was twenty-
nine hours. Louis' patients lived from twenty to twenty-four hours.
Murchison has known death follow in four hours, and not until one hundred and five hours. I have known it from seven to twenty-three hours.
The period subsequent to the accident must needs be influenced by a variety of circumstances, such as the character of the antecedent or coetane-
ous disease, the vital powers of the patient, the extent of the orifice, and
the kind and quantity of lymph thrown out, the part of the bowel, and
the conditions favoring or opposing adhesion. If in a fever of the adynamic type, when the powers of the system are much reduced, the shock
may be such as at once to usher in a fatal collapse. If the opening be in
immediate apposition with another coil of the bowel, a solid organ, or the
walls of the abdomen, the extrusion of the contents of the canal may for
a time be arrested. Bristowe relates a case in which the patient lived
fourteen days after perforation. I remember an instance in enteric fever
in which there was a hole that would have admitted a swan-shot on the
lower part of the ileum, but depositions of pearly lymph had so effectually sealed up the opening that none of the intestinal contents had escaped.
When, however, they do escape, the inflammation becomes so intense that
remedies are generally powerless. Chomel, Louis, Rokitansky, and Jenner
say it is *always* fatal. Tweedie, Todd, Ballard, Fox, Bell, and Murchison aver that they have known recovery. The last-named relates the instance of a girl of sixteen, who, on the thirty-first day of fever, was suddenly seized with severe pain and tension of the abdomen, urgent vomiting, and all the symptoms of collapse. A grain of opium was given every
second hour, and during the first thirty-six hours ten grains were taken.
The patient made a tedious recovery, and was discharged from the hospital fifty-five days after the commencement of the Peritonitis.
    In some exceptional examples, the more formidable symptoms will
apparently subside, and life be preserved for even several days. This
deceptive kind of amendment should not, however, throw the physician
off his guard; he should not forget those grave and alarming indications
which pronounced the existence of the accident, as it almost invariably
proves that the mortal end has only been deferred, not averted. In the
case observed by myself, if there was no absolute escape of the intestinal
contents, the soft lymphic plug could not for any great length of time
have sufficed to act as a barrier to extravasation. Some slight strain, as
in the evacuation of the bowels, coughing, sneezing, or the mere motion
of the body, might doubtless have been sufficient to remove the non-organ-
ized albuminous deposit, and render the opening free. Notwithstanding
the well-nigh hopelessness of all cases in which there is positive solution
of continuity, it is from pathological reasoning a possibility that recovery
may succeed. Nature attempts to repair the lesion by throwing out plastic materials, and if these,—by utter rest, and by opiates subduing the
peristaltic action of the bowels,—be allowed to lie in contact with the
breach sufficiently long to become permeated with new vessels—to be
organizea—the orifice may be repaired: such reparation, however, can
only be effected when the hole is small, and then it is but a mere possibility.

Though the first symptoms of perforation are nearly always distinct and terrible, in exceptional cases they may be ill-defined and obscure; or they may gradually assume increased severity. They will be influenced by the size of the aperture; for instance, the solution of continuity, when it takes place in the appendix, is sometimes very minute, and the escape of irritant matters inconsiderable. The orifice may at first be small and by degrees enlarge, and relatively with the enlargement (and consequent greater extravasation of liquid and fæcal contents) will increase the irritation conferred to the sac and the more manifest phenomena of inflammation. Confirmative of these assertions, Dr. John Harley may be cited. "In some cases," says this physician, "the perforation has taken place so gradually, the aperture formed is so small, and the extravasation so inconsiderable, that the symptoms of Peritonitis come on and attain their maximum very gradually, and without any sudden increase in the severity of the symptoms." [1]

The colon is occasionally perforated in fever, but it is much less prone to this result than the parts last named. Chomel, Brinton, Forget, and Murchison mention five instances. In two out of these cases the opening was at the junction of the transverse and descending colon; and in the three others at the junction of the sigmoid flexure with the rectum. [2] The last-named authority lately gave a good example of the giving way of the large intestine. "A young man of eighteen was admitted into the Fever Hospital, Aug. 23, 1865; he had been ill fourteen days, and on admission was very ill of typhoid fever with Peritonitis. The pulse was quick and feeble, the body enormously distended and tender, the motions frequent and watery, and the breathing thoracic. He died Sept. 7. Inspection discovered the entire surface of the peritoneum to be coated with a thin layer of lymph which could be stripped off with a knife. There were three perforations in the large intestine, one about three and a half inches below the valve, and two in the sigmoid flexure. There were no contents of the bowel in the serous sac." [3]

With respect to the average of perforation in fever, Murchison states that out of 435 autopsies recorded by Bretonneau, Chomel, Montault, Forget, Waters, Jenner, Bristowe, and those made at the London Fever Hospital, it occurred in 60 cases, or in 13·8 per cent. [4] It probably happens in about three per cent. of those who have enteric fever, and more frequently amongst males than females.

In chronic dysentery, sometimes, after ulceration has destroyed the mucous and muscular coats, the peritoneum is penetrated. In such instances the special and general symptoms, which characterize the primary disease, point to a correct diagnosis. In cancer of the bowels perforation may occur: it is more frequent in the large than small intestines, and Rokitansky says the colon is almost exclusively the seat of· cancerous degeneration. I saw in consultation some time ago a gentleman laboring under diffuse Peritonitis, which had evidently been caused by a large hard tumor, the size of a cricket-ball, in the left hypogastric region. The stools were flattened, but the passage was evidently quite patulous. I gave it as my opinion that it was a case of cancer of the large bowel. A surgeon was at this juncture called in, and he strangely enough proposed Amussat's operation merely to give exit to the flatus, when large pieces of fæcal matter were voided, but fortunately that suggestion was

---

[1] System of Medicine, vol. i. p. 570.
[2] British Medical Journal, Dec. 2, 1865.
[3] Murchison on Fever, p. 551.
[4] On Fever, p. 511.

overruled by two of the most eminent members of the profession. In the course of a few days the patient died. Perforation was announced by a sudden and terrible increase of pain, small pulse, sunken features, and cold extremities. The autopsy revealed abundant proofs of foregoing and present Peritonitis. There were several pints of serum in the abdomen, which contained loose flocculi; the descending colon was adherent to the abdominal walls, and a little above the sigmoid flexure was a cleanly cut, punched hole, the size of a small pea, through which a large quantity of thin feculent matter had passed into the peritoneal sac. The upper third of the rectum, and the opening into the sigmoid flexure, were the seats of cancerous deposit, and the canal was patulous.

Habershon divides perforations into two great classes, those which arise from disease commencing in the intestine itself, as by the ulceration of fever, dysentery, cancer, and the various forms of insuperable constipation; and those in which perforation is from without, as in strumous Peritonitis, ulceration of the stomach extending to the transverse colon, hydatids, and abscess of the liver, calculi, abscess in the other solid viscera or abdominal walls, cancer, extra-uterine fœtation, and external injuries.[1] It may be caused by laceration of the gall-bladder. Barthez and Rilliet mention a case in a girl of twelve whilst in fever, and Murchison gives another instance in a young man of nineteen, who was suddenly seized with Peritonitis on the fifteenth day of the fever, and who died in twenty-six hours. It is rarely observed as the result of tubercle. Sir Thomas Watson, in his large experience, only remembers a single instance. Of fifty-six cases collected by Habershon, four only were from strumous disease. Jenner once knew a softened mesenteric gland give way during fever, and Buchanan saw a fatal case of Peritonitis from the bursting of a softened embolic deposit in the spleen of a typhous patient.

Puerperal Peritonitis.—In the discussion of this part of the subject I may here observe that it is not my purpose to enter upon the consideration of puerperal Peritonitis as it occurs epidemically; but as I believe with many other writers that puerperal women are liable to a simple form of Peritonitis, its description necessarily comes within the limit of this article. Sporadic cases from time to time occur without the diffusion of the disease, but even then it is right to observe the utmost caution, as so much doubt is always involved with regard to its contagious nature. Inflammation of the serous covering of the uterus and its appendages may, I believe, supervene as an incidental circumstance, without the superaddition of a specific poison. The great effort of the organism, the irritable condition of the body, after the exhaustion of expulsive endeavors, the long distention of the uterus and the abdominal walls, and their sudden contraction; the friction of opposed surfaces in the abdomen during labor, and the great excitation given to the circulatory and nervous systems, may produce Peritonitis. Other causes operate in the production of this result, such as injuries inflicted during instrumental delivery, in turning, adhesion of the placenta, the use of cold affusions in flooding, and the improper administration of stimulants. Contamination of the blood, originating in the body itself, without reference to external agencies, as when absorption takes place from putrid coagula or a piece of retained placenta, is another mode by which the malady is originated. In uræmic poisoning, as before remarked, the serous membranes are pre-

---

[1] Diseases of the Abdomen, 2d edit. p. 530.

disposed to inflammation, and the blood vitiation during parturition resembles this cause.

There is, I need scarcely say, still much conflict of opinion relative to the real nature of abdominal inflammation after child-birth. By some it is yet maintained that Peritonitis and puerperal fever are identical—that these terms express but one affection. It is true that in a large proportion of those who die of puerperal fever the peritoneum is inflamed, but this membrane is not *always* involved; and although this form of inflammation accompanies this disease far more frequently than any other form, yet puerperal fever is something still more. Of 222 autopsies of puerperal fever, given by Tonnelli, in 193 were traces of Peritonitis; in 29, or one-eighth, there were no traces whatever. Of 44 cases examined by Lee, the peritoneum and uterine appendages were inflamed in 32, or in the relative proportion of 8 cases out of every 11. Dr. Bartsch, in a report of the Midwifery Institution at Vienna, records the morbid appearances of 109 cases of those who died of puerperal fever, and in this report puerperal fever is distinguished from Peritonitis and metritis. "The cases of puerperal fever," he says, "occurred *seldom under the form of puerperal Peritonitis,* but generally as inflammation of the uterine veins, giving rise to the production of pus in these vessels, and the general symptoms accompanying its absorption." [1] Let any one, says Fleetwood Churchill, compare a case of simple inflammation of the womb or peritoneum in childbed with a case of epidemic puerperal fever, their symptoms, course, and the effect of remedies, and I do not think a doubt will remain upon his mind, that although the latter is a local disease, it is not exclusively so.[2]

The symptoms common to this form of Peritonitis may come on in a few hours, a few days, or even so long as two or three weeks after delivery. Pains and rigors are generally the first indications, and pain on pressure is more distinctly felt at the hypogastrium than at any other part. The skin is hot, the cheeks are flushed, the pulse is quick, and the respiration hurried. The pain soon radiates from the hypogastrium into the iliac fossæ, and then to the other parts of the abdomen. It is not always severe, and is sometimes characterized by paroxysmal attacks, the patient being free from suffering during the intervals; nor can it be said that this symptom pain is pathognomonic of puerperal Peritonitis, because post partum uterine pain may be urgent when there is no co-existent inflammation, and there may be inflammation with little or no abdominal pain. Churchill asserts that he has seen five or six cases of intense Peritonitis as proved by dissection, in which there was neither pain nor tenderness;[3] and Ferguson records that he has known nineteen cases in which there was no pain.

The abdomen suddenly becomes large, more quickly and to a greater extent than in any other kind of Peritonitis, which may be accounted for by the often relaxed and resistless condition of the muscular system of parturient women, and because the abdominal walls have so recently distended. At the onset of the attack, the uterus can be felt above the pelvic brim, soft, flabby, and uncontracted, but as the distention obtains in greater degree it cannot be distinguished. The lochia are at once diminished or suspended, or their absolute suppression may precede the inflammatory phenomena. If the milk has begun to flow, its secretion

[1] *Lancet,* April 16, 1836.
[2] Diseases of Women. Syd. Soc., p. 35.
[3] Diseases of Women, 5th edit. p. 783.

is arrested; if it has not begun, it is prevented. If the mammæ have been full and rounded, they fall in and are flaccid and smaller. The pulse varies, but it is always above, in the great majority of cases greatly above, the normal standard. In non-inflammatory, uncomplicated cases, the circulation may be accelerated, for a day or two, or two or three days, but there is a gradual declension of its frequency from the time of delivery. If, however, after delivery the pulse shall have fallen to, or near, its natural number, and it then suddenly begins to rise, accompanied by local pain, higher temperature, thirst and diminished secretions, the cause is often obvious.

After-pains may be confounded with those of inflammation. They come on soon after delivery, but decrease in force and frequency as time wears away. Peritonitis does not come on so soon, and its symptoms become more and more proclaimed, instead of diminishing. After-pains are associated with a firmly contracted uterus; Peritonitis with a relaxed uterus. Remedies which relieve the former are useless or harmful to the latter. In the one affection the circulation may be natural; in the other it is never so. At the first the diagnosis is very difficult, because after-pains may be followed by inflammation, and for a time the symptoms be mixed up; but the progress of the case leads to a correct conclusion. When puerperal Peritonitis is accompanied with intestinal irritation and the inflammation has extended to the mucous membrane, sickness and diarrhœa may be urgent. When the malady terminates by resolution, the pain abates, the tympanitis declines, the pulse becomes fuller and slower and softer, the skin cooler and moist, the tongue cleaner, the lochia are re-established, the breasts become rounded and milk begins to flow, the legs can with more comfort be extended, and the patient can lie on her side. The conditions of approaching dissolution are — weak and thready pulse, varying from 120 to 160; the abdomen keeps distended and tender, cold clammy sweats come on, the extremities become cold, the breathing is quick, shallow, and thoracic, she lies on her back with legs drawn up, the features are sunken, and the mind often remains calm and clear to the close.

Perityphlitis.—This particular form of disease has been more fully described by French than British pathologists. MM. Husson and Dance[1] give an excellent account of the affection; and it is also well described by Dupuytren, Menière and Duplay. Amongst the English authors may be named Copland,[2] Syme,[3] Craigie,[4] Farrall,[5] Burne,[6] Sellar,[7] and West.[8] The disease originates in the tunics of the cæcum, and by some it has been named pericæcal abscess; the glands or follicles of this organ at the first become inflamed and then pass into the ulcerative condition. The ulceration of this part of the large bowel may insidiously destroy the mucous membrane, implicate the sub-mucous cellular tissue and peritoneal coat, and either cause inflammation and lymphic adhesion of the latter, or its fatal perforation. When agglutination occurs the lesion may

[1] Mémoire sur quelques Engorgements inflammatoires qui se développent dans la Fosse iliaque droite; Répertoire d'Anatomie, &c., t. iv. p. 74.   Paris, 1827.
[2] Med. Dict. art. Cæcum.
[3] Principles of Surgery.
[4] Pathological Anatomy, 2d edit. p. 632.
[5] *Edinburgh Medical and Surgical Journal,* vol. xxxi. p. 1.   1831.
[6] Medico-Chir. Transact. xx. p. 200, and xxii.
[7] *Northern Journal of Medicine,* July, 1844.
[8] Diseases of Infancy and Childhood, 5th edit. p. 656.   1865.

be arrested. Craigie defines the malady to consist in inflammation and suppuration of the cellular tissue connecting the cæcum to the quadratus lumborum muscle and other parts, or in inflammation and ulceration of the mucous membrane of the cæcum; and Sellar says its pathological seat is in the cellular tissue between the fascia of the iliacus internus and the coats of the cæcum.

The causes of perityphlitis may be referred to the peculiar position of the cæcum, as well as to other circumstances. It is attached to the muscles of the right lumbar region, and its sacculated pouch depends below the ileo-cæcal outlet, and as all physiological anatomists observe, its contents have to be propelled against gravity; and it thus may become distended with fæcal matters, and such irritation be instituted by its distention and pressure as to set up inflammation of the lining membrane. Again, hard and indigestible articles of food, the stones of drupaceous fruits, seeds, pieces of bone, and metallic, porcellanous, and vitreous fragments have been known to give rise to it. The complaint has in several recorded cases been present long before its nature has been discovered. Its earliest conditions are rendered manifest by the tumescence and dulness on percussion at the right iliac fossa. The circumscribed swelling may extend across to the umbilicus, and when such is the case Peritonitis is generally the accompaniment of other pathologic changes. The patient will complain of pain at the upper part of the thigh, and this has not the same freedom of motion as the other limb. It has repeatedly been found that there has been irregular action of the bowels, associated with colicky pains, which radiate from the iliac region. Dr. West says, that in children the bowels are mostly relaxed, and that pain in the stomach is an initiatory symptom; and he also remarks, that the prominence in the right flank sometimes assumes that of an elongated tumor, which reaches from the ramus of the pubis nearly to the hypochondrium, and has a brawny hardness.[1]

When the ailment has for some time subsisted, lymph and purulent matters are deposited in the cellular tissues behind the cæcum, and so long as the strong iliac fascia prevents the escape of pus, a deep and irregular abscess is formed. The secretion at length most frequently passes through the cæcal parietes at the part uncovered by the peritoneum, as recorded by Copland, Duplay, and others. In some instances it is infiltrated into the cellular tissue in front of the iliacus internus, and effects an exit near the anus; or it may pass into the folds of the meso-colon, or make a sinus and be evacuated externally, as in examples related by MM. Husson, Dance, and Menière. Dupuytren knew it extend so high as the right kidney, and so low in the pelvis as to collect between the rectum and bladder. The perityphlitic inflammation may be circumscribed and rather of the sub-acute than the acute type, with adhesion of adjacent surfaces. When the matter perforates the serous sac, diffuse and fatal Peritonitis ensues.

Peritonitis of Children.—Acute Peritonitis seldom occurs in infancy and childhood. It has been more frequently observed in young infants than in children several years older. Some have declared it may affect the fœtus; in all such instances syphilis in the mother has been regarded as the cause, nor is it improbable that a general taint in the mother should impart disease to the child. Irritation of the digestive surface is more common in children than inflammation of the serous tunic. When Peri-

---

[1] Diseases of Infancy and Childhood, 5th edit. p. 657.

13

tonitis does occur, it is generally as a complication or sequel. It may however, be primary as well as secondary; it may be partial or general, acute or sub-acute, and then pass into the chronic condition. When it appears it is mostly after one of the exanthematous fevers; more especially after scarlatina or measles.   Dr. West has not known more than half-a-dozen instances of acute general Peritonitis in childhood.[1]   It has prevailed among young infants when exposed to deleterious external agencies.   According to M. Thore,[2] at the Hospice des Enfants Trouvés, at Paris, six per cent. of the infant mortality was from acute Peritonitis. It usually came as the complication or sequel of some other ailment, and no child above ten weeks was attacked by it.   The fatal end was generally before twenty-four hours.   Of sixty-three inspections in no case was there pus, but in all a greater or less amount of serum on which flocculi floated, and the intestinal coils and solid viscera were adherent.   In seventeen out of the sixty-three, erysipelas had preceded the Peritonitis.   Pleuritic effusion was discovered in a third of the examples.

The usual symptoms are pain in the bowels, which at first resembles common stomach-ache.   It alternately subsides and returns, and there is mostly diarrhœa.   In the course of a few days the pain becomes more fixed, and the child frequently complains of pain in the right side, and if old enough he indicates the locality by putting his hand on the cæcal or umbilical region.   The pyrexial phenomena are proclaimed, the little patient looks haggard, he is restless and continually alters his position; pressure over the part makes him cry, and the abdominal muscles are tense.   He lies on his back, often with legs extended, and the sickness is not so urgent as in the adult.   According to Dr. West, when the affection is of cæcal origin, the right leg is often drawn up and the left extended.

Dr. George Gregory a long time ago described a form of marasmus, which he believed to be primarily disease of the peritoneum, and which he conceived to differ from what Pemberton terms "irritation of the intestines," and the kind of marasmus originating in the mucous membrane.[3] From being met with in scrofulous children, and an "imperfect kind of pus" being produced, he named it scrofulous inflammation of the peritoneum.   He regarded it to be distinguished by abdominal tenderness, shooting pains which at the first come on in paroxysms, but at length increase in frequency and violence.   The pain on touch is first localized, and then becomes diffused.   Inspection revealed pus and agglutination of the viscera.   But the account of this author applies more to chronic than acute Peritonitis.   In acute Peritonitis of children pus is a rare consequence; when it is formed it gravitates into the lower parts of the abdomen, and is deposited in one or more collections or septa.   It may be evacuated by pointing externally, as in empyema, or effect an exit by the bowels, and it is possible recovery may follow, but such is a possibility rather than a probability.   When it occurs consecutively, as after some fever, and when the powers of vitality are lowered, turbid serum with a few floating flocculi is the common product, as I have already observed when speaking of the non-plastic type of the disease.

Complications.—This affection is often complicated with some other disease.   It may be complicated with *gastritis*, a disease which rarely or never

---

[1] Diseases of Infancy and Childhood, 5th edit. p. 654.
[2] De la Péritonite chez les Nouveaux-nées, in the Archives Gén. de Méd. August and September, 1846.
[3] Medico-Chirurg. Trans. vol. xi. p. 263.

occurs in this country as an idiopathic affection, although it is said to do so in warm climates. The physician will, in nearly all cases, discover from the history of the case, or collateral circumstances, the cause of the inflammation. Gastric Peritonitis may be fatal without the contents of the stomach being poured into the serous sac, and without solution of continuity, especially when it occurs in a secondary form. But in such examples the inflammation is only limited. Sometimes tumors press upon the organ and inflame its serous covering, or the inflammatory condition may be there instituted by contiguity, as when neighboring viscera, such as the liver, spleen, and intestines, are thus primarily diseased. Carcinoma, especially of the pyloric end, will sometimes, by the mechanical pressure, give rise to the result in question; when this happens the Peritonitis is generally of the more chronic description. In that form of ulceration of the stomach, which occurs mostly in young women, the general health is often not much affected. It is often in association with chlorosis, amenorrhœa, leucorrhœa, or sub-mammary pain, and the patient is apt to complain of a gnawing sensation at the epigastrium, accompanied with more or less of anorexia and vomiting. When the gastric peritoneum is rent or perforated by ulceration of the inner tunics, the pain is excessive, the powers of life are rapidly subdued, and death is inevitable.

When the peritoneum is inflamed in *hepatitis* it is generally in a partial manner, and it continues to be circumscribed unless extravasation of some description result, which is occasionally the case, and then the entire sac at once assumes the same morbid condition. Inflammation may begin in the parenchymatous structure and extend to the serous coat, and when such is the fact, the pain becomes more acute and defined, and the pyrexial symptoms are more pronounced. The right hypochondriac region is often full and tense, the normal lines of dulness are extended, there is pain on pressure and deep inspiration, and dyspnœa, coughing, and vomiting are frequent accompaniments. The patient cannot lie on his left side, and the recti muscles are rigid. When the convex surface is affected, the diaphragmatic investment assumes the same disease, and cough is a prominent symptom. The convexity may be inflamed without the appearance of jaundice. When the concavity is inflamed the stomach mostly becomes implicated, sickness is urgent, the gall-ducts are more or less obstructed, and jaundice, in greater or less degree, is a common result. When the parenchyma is alone inflamed, the pain is of a dull, aching character. When the serous tunic is involved, the pain is sharp and acute. When lymph in considerable quantity is effused, the organ becomes adherent to adjacent surfaces, and if the albuminous exudation gravitate to the lower part of the abdomen, agglutination of the intestinal folds occurs. When hepatic abscess points to the surface, partial Peritonitis, by pressure, is induced. The effused lymph is protective from the worse consequence of extravasation. Hydatid tumors may, like abscess, excite adhesive inflammation. Cancerous growths occasionally produce sub-acute hepatic Peritonitis, but the symptoms are ill-defined and obscure. And the same remarks apply to the tubercular masses in the capsule of the liver.

Sometimes we observe *acute splenitis* as an intercurrent complaint during the progress of intermittent fever. But, as I have more fully insisted in the article on Diseases of the Spleen, this organ is infinitely more prone to a chronic form of congestion. Sometimes, when during the cold stages the capsule becomes suddenly distended, such tenseness so stretches the fibrous and serous tunics as to usher in the inflammatory process; then pain of sharp and stabbing character, increased by pressure, is felt beneath the left costal cartilages radiating through to the back; the skin is hot, the pulse quick and hard, the urine high colored and scanty, the tongue furred, the bowels are confined, and if the under surface of the diaphragm has become affected, cough and dyspnœa are associated symptoms. The patient lies partly on his back with trunk curved to relax the abdominal muscles. Towards evening there is exacerbation of the symptoms. Post-mortem examination reveals the serous investment thick and reddened, and the organ united to neighboring parts by albuminous exudation; and it is here not unworthy of remark, that in the peritoneal inflammation of this viscus, cartilaginous and ossific conversions are more frequent than in the peritoneal inflammation of the other solid abdominal organs.

In *enteritis*, when all the coats of the bowel are inflamed, the disease may commence in the mucous membrane, at first sickness and purging being urgent. In such cases colicky pains come on at intervals, and moderate pressure produces little or no uneasiness, and at this stage of the malady it is often difficult to form a correct diagnosis. If the complaint make progress, if the skin become hot and dry, the pulse quick, the face flushed, and pain be felt on pressure, it is of great practical importance to distinguish the kind of lesion to which the disease has advanced, because remedies which would relieve the colic would be absolutely injurious in inflammation. Instead of diarrhœa there is often constipation; thus it is when mechanical obstruction of the gut is the cause of its being inflamed, as in intussusception, and when tumors block up the passage, and vomiting of stercoraceous matters proclaims the inverted action of the bowel. The general and special signs of the peritoneum being inflamed are the same as those above described. In *children* the complaint is frequent during dentition, and it sometimes comes on as the sequel in eruptive fevers. Crude and indigestible articles of food in these little patients are often the cause. Its advent is marked by languor and peevishness, the child is restless and complaining, green mucoid stools emitting an offensive odor are voided, the cheeks become flushed, the belly tender, and all the conditions of peritoneal inflammation are superadded to a fever of the remittent type. And dissection sometimes exhibits the entire substance of a portion of the ileum presenting a gangrenous appearance in addition to the ordinary products of serous inflammation.

In *nephritis*—which is in the great majority of instances brought on by calculus in the pelvis of the kidney, blocking up of the ureter, some irritant drug, or some blow or external injury—severe pain over the loins following the course of the ureter on the same side, and, in the male, retraction of the testicle, high-colored urine, and nausea and vomiting are common symptoms; and, as is occasionally the case when ischuria renalis supervenes, uræmic symptoms are apt to mask and obscure the otherwise more apparent features of peritoneal complication (*perinephritis*). The urinary bladder may be acutely inflamed (*cystitis*), the inflammation originating in the mucous membrane, and being extended to the muscular and serous coverings. It is caused by calculi, irritant drugs, retention, surgical operations, and external injuries, and the Peritonitis may be partial or general.

Hystitis is very rarely observed in the unimpregnated uterus; it may come on after menorrhagia by sudden suppression of the catamenia, long walks, wet and cold, and I have known it induced by the incautious use of topical applications. It is most frequent after delivery, and the fundus is the part mostly first affected. When the peritoneal investment becomes implicated the disease often assumes an alarming character. *Ovaritis*

may be presented in one or both the ovaries without the uterus being inflamed; in the larger number of examples, however, it is the complication
of general Peritonitis, or antecedent uterine inflammation. Deep-seated
pain in one or both of the pelvic cavities indicates the lesion, and when
the peritoneum is affected the pain becomes exceedingly acute, and an
aching, wearying sensation extends down into the groins and thighs.
There is often frequent desire to micturate, and when the disease is continued to the posterior portion of the peritoneum the rectum is rendered
irritable, and there is constant inclination to evacuate the bowels. Puffi-
ness or swelling is sometimes seen over the ovarian region, and that part
is most painful on the least pressure, and the sickness and vomiting are
often distressing.

The comparatively recent establishment of that great surgical operation *ovariotomy*, more especially as practised in this country, has proved
that the peritoneal sac can be laid open, and its inner surface exposed
over a great extent, and for a considerable time, without the production
of such fatal results as it was formerly believed would inevitably follow.
It now appears, from a large accumulation of cases, that in a healthy subject, and especially in the unilocular tumor, and when there are no attachments, the peritoneum may be cut, and freely, without the consequent
inflammation being always formidable.

There are some other affections with which Peritonitis is occasionally
complicated. In pericarditis and pleuro-pneumonia it sometimes happens
that the inflammation spreads to the peritoneum: but in such instances it
is often extremely probable that a contaminated state of the circulatory
fluids constitutes the predisposing cause, and that the irritation existent
in one of the great cavities is readily transferred to another, and that an
adjacent membrane of similar structure, and under general predisponent
circumstances, will take on the same morbid action. And, conversely, we
know that Peritonitis often extends to the pleura, and it is not uncommon,
as I have lately seen, to find hepatitis associated with dulness, moist crep-
itation, and all the other physical signs significant of inflammation in the
lower third of the right thorax; and when the spleen is greatly enlarged,
or in acute splenitis, the same conditions obtain at the base of the left
lung; pressure and the proximity of like structures being the cause of
such extension. In empyema the diaphragm may be rendered convex towards the abdomen, pushing down the abdominal organs, and friction and
pressure induce Peritonitis; and in the enlargement of the liver or spleen,
or an encysted kidney, or an ovarian tumor, this partition may be thrust
up so abnormally into the chest as to press upon and excite the pleuro-
pulmonary tissues to active inflammation.

MORBID ANATOMY.—The morbid appearances of Peritonitis are very
various, being modified by a number of circumstances; such as the type,
the primary or secondary character of the attack, the condition of the
blood, the amount and kind of disease in the viscera, and more especially
of the solid organs.

Before speaking of inflammatory change, it may be observed that
serous membranes may be simply congested, presenting a condition analogous but not amounting to inflammation, and this hyperæmic state may
be transient, temporary, or long-continued. When often returning or for
some time existent it may give rise to excess of secretion, which is chiefly
serous; nevertheless it may contain some coagulable matters, but their
amount will be dependent upon the increase or diminution of the fibrinous
and albuminous constituents in the blood. Such abnormal afflux of blood

to this membrane may subside spontaneously, or there may be hæmorrhage into the sac, and such hæmorrhage may be passive or active,—it may be by transudation or rupture. Exhalation into the peritoneal cavity some times occurs, when a sanguinolent serum and an injected membrane are discovered. In visceral laceration considerable collections of blood of course may follow.

The gases generated in the cavity of the peritoneum are sometimes in great amount; they are in nearly all instances the result of cadaveric change and the decomposition of the secretions. In empyema, gases are produced when there is no solution of continuity in the pleura, and the same may result when there is pus in the abdomen and the peritoneum has maintained its integrity; but they may have their origin in ulceration of the intestines, or traumatic injury.

The first inflammatory change in the peritoneum is the loss of transparency and of that shining polished appearance proper to its healthy structure. This dulness or opacity is accompanied by diminution of the lubricating secretion, and Baillie, Bichat, and Knox affirm that the membrane becomes dry. But such dryness is more apparent than real, because when handled it feels moist and unctuous. The sub-serous vessels become injected, and may be seen through the fine membrane in hair-like streaks, arborescent and ramified, or in a confused net-work, and when much crowded a velvety appearance is imparted. The degree or shade of redness depends upon the period of congestion, the kind of inflammation, and the condition of the blood. When the hyperæmia has for some time continued, or in sthenic inflammation, the hue is light red; when the congestion is but recent, or the inflammation of asthenic type, the color is less vivid and may be darker and venoid.

With the progress of the disease, vessels in the membrane which were colorless enlarge so as to admit red-blood globules. At various points small sub-serous sanguineous effusions are seen in the shape of bloody puncta; sometimes these are so numerous as to exhibit a spotted or speckled appearance, or they may coalesce and form red configurated patches of various sizes. I have said that at the first there is diminution of the lubricating fluid. In the course of a short time (at periods differing according to certain conditions which obtain, such as the mildness or severity of the attack, the general powers of the system, and the like) this secretion is re-established, and if the malady end in resolution it manifests all the characteristics of the natural state; but if the complaint progress it is augmented in quantity and altered in quality. The free surface of the peritoneum is then bathed with a semi-transparent homogeneous fluid, and the sub-peritoneal tissue is surcharged with a sero-albuminous secretion, and frequently the peritoneum proper can be stripped off with undue facility. This infiltration, however, at length permeates the serous tunic, when it and the filamentous layer become so confounded that it is not easy to trace the line of union. Under such circumstances the membrane is not only rendered opaque, but it looks thick and tumefied, and if carefully examined it feels rough, has lost its lubricity, and close inspection detects a viscid albuminous deposit varying in thickness according to the duration and severity of the attack.

The new or morbid secretion which is effused soon separates into two distinct forms,—a thin and watery whey-like fluid, and a thick gelatinous, pulpy, or more solid portion; the former constituting serum, the latter coagulable lymph, or, as it is otherwise named, albuminous exudation or plasma. The relative proportions of the fluid and more solid parts vary in each individual instance. Sometimes we find no serum whatever, and sometimes the effusion consists almost entirely of serum, the only traces of the albuminous exudate being minute flocculi floating in the fluid and rendering it turbid. In the inflammation of metastasis and low types of Peritonitis the effusion is sometimes puriform or absolutely purulent. In acute sthenic Peritonitis the lymphic deposit is great. It is thrown down on the free surface of the sac in various amounts according to the condition of the circulation, and the violence of the inflammation. It may be a mere film, or in a layer several lines in thickness. It differs in color, being sometimes of a grayish red, but is more frequently of a yellowish straw color. When abundant, it lies in smooth or corrugated plates; it is also found in honeycomb arrangement, in bands or bridles constituting bonds of union of varying thickness uniting the viscera, or it may be encircling the gut; it is generally seen in masses filling up the interspaces, and when lying between the intestinal folds it assumes an ill-defined prismatic configuration. The viscera are not only glued and matted together, but there is mostly more or less of adhesion to the parietal peritoneum. When a portion of the adventitious stratum is detached from the peritoneum, the coherent surface of the new product exhibits an irregular villous character, and it is speckled with small bloody puncta produced by torn capillaries, and the sub-serous tissue is ecchymosed. The new formation being at first villous, becomes smooth and more dense, and at length assumes a structure and qualities analogous to the true peritoneum.

If the exudation be submitted to the microscope new vessels are seen to permeate its substance, and more especially in the central portions. That they are connections or prolongations of the peritoneal capillaries is beyond dispute, although we cannot always trace their continuous structure. It was believed by Hodgkin[1] that new vascular extensions are carried out into the exudation, and that subsequently towards the peritoneum they contract and become nearly or quite invisible. This author is of opinion that the delicate parietes of the extreme vessels give way, that minute quantities of blood are received into the exudation, and that such are the first beginnings of those minute cavities which are destined to become vascular.

It is quite evident that the plastic effusion is an irritant to the serous surface, because when deposited on one part of the peritoneum, and any other opposing part comes in contact with it, such readily takes on the inflamed condition; hence it becomes explicable, in one way at least, why Peritonitis is so liable to diffusion. According to the time which elapses after its production, and the vital powers of the organism, is the degree or completeness of the organization. From being a semi-fluid gelatinous substance it becomes more dense and solidified, the capillaries are more numerous, it contracts in bulk, its filamentous texture is more defined, and it enters into firmer and more intimate union with the organs or parts it covers or connects. Where there is much motion, it is sometimes disposed in a stringy or reticulated manner, and meshes are formed, filled with transparent fluid. Another morbid condition associated with these false membranes is that of serum or sero-purulent fluid being collected between the peritoneum and the false formation, until the latter is raised up and loosened from its attachments and set free in the sac. When these adventitious membranes remain firm and adherent, the original serous membrane beneath them disappears, and their surface assumes the character-

---

[1] Lectures on Serous and Mucous Membranes.

istics of a veritable serous membrane, and it is difficult to distinguish the new from the old. The former secretes a lubricating serum, is influenced by the same kinds of irritation, is liable to become inflamed, and in its turn to throw out true inflammatory products.

The attachments effected by these formations may subsist through the remainder of life. They may be protective and conservative. In the suppurative stages, when abscess forms in the solid viscera, this adhesive inflammation is the method which nature observes for the harmless exit of pus. These bonds of union may continue with little or no inconvenience. By the lapse of time they become thin and contracted, and when health is re-established and the absorbents are active, they may partly or wholly disappear. Absorption begins with the subsidence of the inflammation, and, as Rokitansky[1] remarks, it must, as a matter of course, be influenced by the thickness, that is to say the permeability, of the deposit.

Before the time of the two Hunters it was not by pathologists generally allowed that serous membranes secreted pus without solution of continuity; in other words, without the presence of ulceration. Since then this fact has been universally acknowledged. It may be secreted from the inflamed peritoneum, or from the surface of those adventitious membranes which are formed in the cavity. William Hunter says it is generally thinner than that of an abscess, and the containing surface is more or less covered with a glutinous concretion or slough of the same color as the fluid, in some parts adhering very loosely, in others so firmly that it can hardly be rubbed off, but still the surface covered with these sloughs is without ulceration or loss of substance.[2] Dupuytren and Villerme believe that the false membranes are concrete pus, and Rokitansky is of opinion that pus, under some inherent peculiarity, is a degeneration of plastic exudation. It is more frequently seen in the asthenic, sub-acute, and lower types of the complaint than in the sthenic. In the inflammation of metastasis, when the blood is contaminated, in parturient women, and in children, it is most common. The fluid may be puriform, purulent, or sanious. It may be yellowish green, or brown, or reddish. The peritoneum and sub-peritoneal tissue are much injected, and there is usually great infiltration of the tissues. In some instances it appears as if exuding from the entire inner surface of the peritoneum; in other cases it is associated with adhesions, and is discovered in distinct collections, bounded by organized septa, and resembling separate abscesses. It may be evacuated by ulcerative absorption through the abdominal parietes; by the same process it may pass into the digestive tube, the bladder, or vagina, or through the diaphragm into the thoracic cavity, or effect an entrance into the bronchi, or it may find a way of escape by the psoas muscle.

The pressure exerted by purulent collections is doubtless the main cause of ulceration commencing, but Craigie believes that in these cases sometimes ulceration may result without pressure, being merely the direct and obvious effect of inflammation. My colleague at the Tunbridge Wells Infirmary, Mr. Marsack, made (Sept. 18, 1865) an autopsy on the body of a young woman, on whom he had six weeks previously performed ovariotomy. The coils of the ileum were welded together, and joined to the abdominal walls by organized adhesions. Between the layers of the great omentum were small independent abscesses of creamy pus. In the lumbar region was a bounded abscess-like collection which contained half a pint

<hr>

[1] Pathological Anat., Syd. Soc.
[2] Medical Inquiries and Observations, vol. ii. p. 61.

of pus.   At the sigmoid flexure ulcerative perforation was discovered.[1]
Pressure, caused by a collection of purulent fluid, had been followed by
ulcerative absorption of the tunics of the large bowel.   When this secretion is effused in small quantity it may be absorbed, but if in large quantity and without opening, irritative fever is induced, the symptoms of
pyæmia supervene, and it is then uniformly fatal.   Sometimes adhesive
inflammation in Peritonitis gives rise to very peculiar pathological conditions.   The stomach and transverse colon have, in several instances, been
glued together, and ulcerative absorption has effected a communication
between them, so that the fæcal contents of the large bowel have passed
into the gastric cavity, and thence been expelled by vomiting.   Two or
more coils of the ileum may be soldered together, and an intercommuni-
cating passage established in the same manner.   In such examples the
disease has generally become chronic.

   In the partial or localized forms of acute Peritonitis, when some foregoing visceral disease has extended through to the serous coat, and instituted inflammation in that tunic, we not infrequently see circumscribed
depositions of lymph cementing neighboring parts together while the
remaining extent of the peritoneum is perfectly healthy.   In hepatitis,
when the convex surface is inflamed, strong adhesion is sometimes discovered.   The spleen is in like manner united to the concave surface of the
diaphragm, and the accretion may have assumed a cartilaginous or ossific
character, the latter conversion being in that situation more frequently
seen than in any other part of the abdomen.   In simple ulceration of the
stomach sometimes adhesive ulceration averts a fatal catastrophe by
agglutination to one of the solid organs, or, as it has been repeatedly witnessed, by the production of an aperture into the colon, or sometimes
into the duodenum; and, in a few rare instances, a canulous opening has
been spontaneously made through the abdominal parietes, forming a gastric fistula.   In malignant disease of this organ, most frequently seen at
the pyloric end, there is much soldering together of the adjacent parts;
the peritoneum is opaque and vascular, and the sub-serous tissue is
greatly injected and infiltrated not only with carcinomatous deposit, but
also with serous fluid.   The duodenum, as before remarked, occasionally
exhibits partial Peritonitis from rupture, consequent upon ulceration of
the mucous and muscular coats, as the result of extensive burns, but its
serous investment is more frequently inflamed from the irritation and
pressure resulting from cancer of the head of the pancreas.   When the
jejunum is found morbid it is almost always in connection with the lesion
of other organs.   With regard to the ileum, what has above been said
relative to the perforation of its peritoneal covering was descriptive of its
morbid appearances.   In phthisis sometimes protracted colliquative diarrhœa gives rise to ulceration in its mucous surface, but perforation in
phthisis is exceedingly rare; it is, however, in this complaint occasionally
beheld on or near the vermiform appendix.   In chronic dysentery the
colon may give way, and in such instances there is great destruction of
the other tunics proper to the bowel.   Such examples occur in those who
have died after long residence in tropical climates, and in association
with some form of hepatic disease — very generally with abscess of the
liver.

   In puerperal Peritonitis, according to Dr. Lee, the appearances of
inflammation are sometimes confined to the uterus, but they are much

---

[1] Mr. Marsack's Hosp. Case Book.

more generally extended to other organs. The lymph is mostly thrown out in thicker masses upon the uterus than in any other situation, and this viscus seems to suffer in the greatest degree. In the sub-serous cellular tissue serum and pus are often deposited. The cellular tissue surrounding the vessels of the uterus where they enter and quit the organ, and that connecting the muscular fibres, is often surcharged with serum and purulent fluid.[1] The peritoneum becomes thick and vascular, more especially where it invests the uterus and pelvic viscera, and sometimes, when the malady is intense, the serum is mixed with blood, and pus is found in the pelvis. When death has rapidly followed, the lymphic exudate is semi-fluid, or the surfaces which have become agglutinated are readily torn asunder. The Fallopian tubes and ovaries are sometimes filled with pus or blood.

In the Peritonitis of children the abdominal viscera are found matted together and adherent to the abdominal walls. In some cases the viscera are covered with a thin grayish opaque covering, which feels soft and unctuous, and a turbid, reddish serum in which small flocculi are floating is effused in varying quantity. In that strumous affection which, according to Gregory, gives rise to Peritonitis, pus is secreted. And this physician asserts that sometimes the abdominal cavity will be abolished, the viscera being united in one mass, and everywhere adherent to the parietal peritoneum, the latter in all its duplications being thickened, and the soldered intestinal convolutions inter-communicating.[2] When the peritoneum becomes inflamed consecutively after scarlet fever, measles, rheumatism, or some other fever, an excess of serous effusion is discovered, the albuminous portion being inconsiderable or almost absent. The fluid is of whitish straw-color or of dirtyish red.

DIAGNOSIS.—The more severe forms of acute Peritonitis are fully expressed, and the disease cannot well be mistaken; but in the sub-acute and more partial descriptions, when the disease is not a primary but secondary complaint, or a complication, it may be so masked, mixed up, and confounded with the symptoms of other morbid changes as to render the diagnosis very difficult. In all instances the physician should pay marked attention to the history of the case, as well as to the objective and subjective symptoms, because there are affections which when superficially reviewed simulate this complaint, and it has not infrequently happened that the ignorant or off-hand practitioner has fallen into grave error. The diseases which it most resembles are gastritis, enteritis, colic, rheumatism, neuralgia, hysteria, obstruction of the gall-ducts, renal calculus, and lead-poisoning. With respect to *gastritis*, it is in this country, as I have before observed, rarely or never met with as a purely idiopathic affection. Abercrombie means by this term inflammation of the mucous membrane, and it is in such sense that it is now employed. When the mucous coat takes on this morbid state there may be pain on deep pressure, the sickness is urgent, the thirst distressing, and fluids are constantly ejected. It can almost always be traced to some exciting cause. In Peritonitis there is more difficulty in the etiological conclusion, and in the latter the pulse is smaller and more wiry. The inflammation may commence in the digestive surface and extend to the peritoneal investment, and it then, of course, becomes partial Peritonitis. It occasionally occurs when the gastric portion of the peritoneum is roughened by lym-

---

[1] More Important Diseases of Women, p. 24.
[2] Medico-Chirurg. Transactions, vol. xi. p. 266.

phic exudations that auscultation can detect some friction sound; but this, however, is seldom heard. In the great majority of cases gastritis is referrible to acrid and corrosive poisons. Haller knew it produced by the patient having taken cold water when he was heated. It is frequently very difficult, often absolutely impossible, to diagnose Peritonitis from *enteritis.* Inflammation may begin in the mucous membrane and implicate the peritoneum, or Peritonitis may at length involve all the coats of the bowel, when both diseases obtain. The vomiting is more urgent in enteritis, the bowels are often obstinately obstructed, and gangrene is sometimes the result. The pulse is of better volume than in Peritonitis, and as the rule the patient does not complain of so much pain. In Peritonitis, partly owing to the involution of the parietal peritoneum, the pain on pressure is more acute and superficial, the patient is more averse from motion, the respiration is more thoracic, and the features are more collapsed.

In *colic,* which may be from simple flatulence, the pain and distention may be severe, and even the face may be an index of suffering. When there is very great distention pressure may increase the pain, but more commonly pressure relieves rather than augments it; the circulation is little if at all affected, and there is no symptomatic fever. Frequently constipation and vomiting are associated with other symptoms; the patient complains of a twisting, wringing pain at the umbilicus, which comes on paroxysmally, and there are intervals when the suffering is inconsiderable or absent. This condition of colic is, when regarded alone and as simple colic, not an important affection, but it sometimes comes on as the herald of a more grave disease, and ends by the development of inflammatory symptoms. In *colica pictonum* there is no apparent obstruction of the bowels, although there are the common symptoms of ordinary colic. There are constipation and abdominal pain, even violent pain—*dolor atrox*—but there are other symptoms, such as pain in the head and limbs, a blue, leaden line in the gums, and loss of power in the hands and fore-arms, and the patient is either a painter, or investigation discovers that he has in some way been subjected to lead poisoning. The abdominal muscles in *rheumatism* sometimes are rendered so excessively painful that moderate pressure causes great suffering, and notwithstanding that examples are occasionally observed in which acute Peritonitis has thus supervened, yet such instances are very exceptional, and ordinary observation will generally prevent any mistake in diagnosis. Negative facts will be our chief guide. In such cases the circulation is little affected, the pulse is large and full but not frequent, sickness and vomiting are not present, the countenance has not the pinched, anxious expression which it assumes when the peritoneum is inflamed, and if the abdomen be carefully examined the tenderness will be found more severe at the origins and insertions of the muscles; lastly, it will be shown upon inquiry and examination that rheumatism has recently obtained, or that its symptoms are still present in other parts of the body.

Neuralgia is another affection which mimics Peritonitis. The pain is described as a tight girdle or ligature passing round the body, and imparting a feeling of constriction; it traverses the course of the genito-crural nerve, percussion on the spinal processes detects some tenderness, and the legs and genito-urinary organs are often more or less affected; again, there is the absence of tympanites, pain on pressure, quick pulse, facial collapse, and other phenomena so expressive of Peritonitis, and which I have in detail described above. In that protean malady *hysteria,* which mocks this as it simulates so many other affections, the patient is apt to complain of increased pain almost before the hand has really touched the abdomen and when it does touch it, the pressure does not, as in Peritonitis, augment it. The pulse is natural, the tongue clean, and the countenance does not bear the impress of severe and acute disease. The breathing is not thoracic, the legs can be extended, the decubitus is not dorsal, and borborygmi and intestinal flatulence are often present; again, upon inquiry, it will not infrequently be found that large quantities of pale or colorless urine have been voided, that the uterine functions are at fault, or that some ill-defined spinal symptoms obtain. A comparison of the leading features common to the two affections will leave but little doubt as to the true nature of the ailment.

In *obstruction of the gall-ducts* from calculi, inspissated gall, tumors, spasm, and other causes, the pain is paroxysmal, often excruciating; and with the passage of the obstructing body, and the restored patency of the canal, the suffering at once subsides. There is no pyrexia, the heart's action is little or not at all accelerated, nor is there distention or abdominal tenderness. In addition to such negative there are positive facts; the symptoms of biliary disturbance are mostly present, the alvine dejections are often light-colored, the urine is dark and porter-like, the conjunctivæ are yellow, the skin is tawny, and the pain is localized beneath the margin of the right false ribs. In *renal calculus* the pain radiates from the back round to the abdomen, it comes on suddenly, courses down the direction of the ureters, in the male produces retraction of the testicle of the same side, and shoots down the thigh, when for a shorter or longer interval it declines or entirely subsides, and bloody urine is a common accompaniment.

In puerperal Peritonitis the *after-pains* are associated with contracted, not relaxed uterus, which is the fact in Peritonitis; they gradually diminish, and in thirty or forty hours have become much less in force and frequency. Inflammation of the peritoneum commences at the ordinary date of the after-pains' decline. The remedial agents which relieve hysteralgia do not arrest acute Peritonitis. *Ephemeral fever* is distinguished by its brevity, its milder aspect, by the mammæ remaining of normal size, and those serious conditions which mark the advent of an inflamed peritoneum are wanting. Lastly, in speaking of the diagnosis of this affection, it must be borne in mind that under grave cerebral disease, when nervous sensibility is obtunded, the peritonitic symptoms may be rendered very obscure, and under such conditions diagnosis may be impossible.

PROGNOSIS.—The opinion to be arrived at relative to the result of this disease will be modified and determined by a variety of considerations, and in every case a different array of facts will be presented, all the bearings of which should be carefully scanned. The asthenic is less auspicious than the sthenic type, and when it is the inflammation of metastasis the chances of recovery are less. In *unfavorable* cases, in despite of the best-ordered means of treatment, there is a progressive aggravation of all the cardinal symptoms; the pain does not decline, nor do the distention and the tenderness abate; the breathing is more hurried, shallower, and entirely thoracic, the pulse becomes thready and intermittent, the sickness is excessive, the bowels are generally confined, distressing singultus supervenes, the surface becomes cool, is clammy and relaxed, the legs and feet are cold, the patient falls down in bed with knees drawn up, lies on his back, the Hippocratic countenance is more marked, and often the mind is clear to the end. He sinks by asthenia. In those instances when we can

prognosticate a *favorable* termination, there is remission of pain and ten
derness, decline of the distention, the sickness comes on at longer intervals, and at length abates; the pulse is slower and fuller, the temperature
of the body equable and warm, the respiration is not so quick, and the
diaphragm descends lower down, and the patient can turn on his side.
When we have reason to believe that there is perforation of the bowel,
rupture of the liver or spleen, the urinary or gall-bladder; when we suspect the evacuation of an abscess or the effusion of blood, our prognosis
must be unfavorable, and recovery under such conditions is well-nigh
hopeless. In the consecutive form, when the strength has been undermined by a previous malady, the probabilities of a fatal issue are great.
In puerperal Peritonitis antecedent hæmorrhage and the amount of exhaustion induced by parturient efforts would influence our decision.

TREATMENT.—In every example of acute peritoneal inflammation, the
remedies should be prescribed with a just reference to the emergencies of
each particular case, because no trite and exact rules can be given admissible of universal application. The date of the disease, the powers of the
patient, the kind of pathologic action going on, and the antecedent circumstances so far as they can be ascertained, in conjunction with other
facts, must needs modify our resources, and be suggestive in the selection
of those agents which are accounted as the most effective auxiliaries in
combating the affection. That this disease, like many other ailments,
when seen at the outset, and treated according to science and experience,
can be guided and carried to a successful termination is of such every-day
proof as not to require being insisted upon here. And on the other hand,
if its progress be unrestrained by ignorance or timidity, it soon passes beyond the control of the most vigorous handling and the nicest skill. It is
eminently one of those complaints which does not admit of vacillation and
delay, promptitude and decision of purpose being of paramount importance.

In an acute attack of inflammation of the sthenic type, in the strong
and hitherto healthy, and especially those who have lived in the pure air
of the country, our best ally is *blood-letting;* but it is by far the most successful when performed at the commencement of the malady—as soon as
possible after the pulse has become hard and quick, the pain urgent, and
the disease established. It is then, by making a decided impression upon
the circulating organs, that there is the greatest chance of the inflammatory action being cut short, and of those morbid processes being arrested
which so quickly follow the development of the affection. Nor should we
be deterred from the use of the lancet by the mere *smallness* of the pulse,
because it may feel constricted, hard, sharp, wiry under the finger, for with
the free emission of blood it will increase in volume and become soft and
more natural to the touch. Many authorities, and some of high reputation,
have spoken of the number of ounces which ought to be drawn at a first,
second, or even third depletion, but there is no just rule as regards quantity. One patient will bear a much greater loss of blood than another,
even when the two cases seem to bear a close resemblance. Our real and
only reliable guide must be the effect produced by the abstraction. An
influence must be made upon the heart's action, and the patient should, if
possible, be bled in the erect position. Abercrombie recommends one or
two small bleedings at short intervals after the first in order to keep up the
good results of the primary depletion. There is no doubt if ten or a
dozen hours are allowed to elapse after the first use of the lancet, and before a second visit, that in such long interval the pulse may recover its
strength, the initiatory symptoms in full force return, and a larger quan-

tity of blood will require to be lost. In a disease so perilous the patient should at the outset be seen every two or three, or at least every three or four hours. It is within the first twenty-four hours that blood-letting is of the most avail. When effusion has set in and progressed to some extent, blood-letting is more likely to be harmful than useful. In the young and the robust, in those of ruddy complexion and high arterial action, and those who live in the purer air of the country, bleeding is much better borne, and it may need to be repeated. The dwellers in urban communities, especially amongst the badly nourished and ill clad, such as present themselves at the hospitals of the metropolitan cities and large towns, very rarely, if ever, require general blood-letting, and when it is had recourse to, a smaller quantity is followed by the desired effect.

After the lancet has been used it is excellent practice to follow it up by *local depletion.* Cupping is of course, from the pressure it would give, inapplicable; but twenty, thirty, or even forty leeches at one time may be applied to the abdomen, and often with the greatest benefit. Fomentations, by means of flannels immersed in hot water, and wrung out as dry as possible, the heat and moisture being kept up by their being covered with a large piece of oiled silk, is good treatment, and the flow of blood can thus for some time be promoted; or a large linseed-meal-and-bread poultice, or a bran poultice, produces a soothing effect. In the use of these applications, however, care should be taken to constantly renew them before they become cool, and when they are discontinued a dry hot flannel of three or four folds should be placed upon the abdomen. Another very valuable mode of treatment at this juncture is the employment of terebinthinate epithems. Two or three dessert-spoonfuls of the spirits of turpentine may be sprinkled over the wet flannel, or a large piece of spongio-piline the size of the abdomen may be wrung out of hot water, and the turpentine in like manner sprinkled over it; and these may be repeated two or three times if the patient can endure the applications. I can bear testimony to the very excellent effects of the external use of turpentine, which I have very frequently in this mode recommended, and I believe it to be a most valuable remedy.

The late Dr. Sutton of Greenwich was the advocate of cold applications in abdominal inflammation. He used cold enemata, and cold cloths made wet with evaporating lotions, and, as he asserted, with great benefit. Abercrombie also recommends this method of treatment. " In a considerable number of cases," says this physician, " I have used with evident advantage the application of cold by covering the abdomen with cloths wet with vinegar and water, or even iced water. Injections of iced water have been proposed, and I think it probable might be used with advantage." [1] M. Smoler of Prague has recommended cold compresses often renewed, and laid on the abdomen, their application being desisted from as soon as the patient sleeps; but he never allows the patient to change them with his own hands.[2] Not having any personal experience of cold appliances, I shall therefore not do more than mention a remedy to the success or otherwise of which I can bear no testimony. It would to myself at least seem of doubtful utility in many cases, and one involving great risk in others, and I prefer what I believe to be equally efficacious, and certainly safer, namely, warm fomentations.

After the abstraction of blood a large dose of *opium* should at once

---

[1] Pathological and Practical Researches, 3d edit. p. 173.
[2] Betz's Memorabilien, and Gaz. Med. Lyon, Nov. 16, 1865.

be administered, and two or three grains may be given in urgent cases. It then not infrequently happens that the patient has a tranquil sleep, after which he awakes with less pain, a moister skin, and with remission of the symptoms generally. In those instances in which sickness and vomiting from time to time come on, opium often acts more beneficially. If we wish to influence the system by mercurials, one grain of opium and three grains of calomel may be taken every four or six hours, and mercurial frictions on the thighs and in the axillæ can at the same time be adopted by means of the linimentum hydrargyri, which is perhaps the most convenient preparation for this purpose; or two grains of calomel and half a grain of opium may be given every second hour, and the inunction being also used until some slight effect be produced on the gums. Another mode of administering opium, especially when the stomach is irritable and ingesta are rejected, is by enemata. Thirty or forty drops of laudanum can be injected in two or three ounces of starch gruel, and such repeated according to the exigencies of the case. If the bowels should be loose and the rectum inclined to expel its contents, a suppository, composed of a couple of grains of solid opium with a sufficient quantity of Castile soap or cocoa-nut butter to form a conical mass, may be introduced *per anum*, and such from time to time as the physician may deem desirable. The indications denoting benefit having accrued from the above-named remedies will be mitigation of pain, softer and fuller pulse, easier and slower breathing, more relaxed skin, and diminution of the abdominal distention; the face, too, will look calmer and more natural, and the patient probably give expression to a more comfortable feeling.

Vesication is another of our aids in guiding the malady to a favorable issue. It may be done by means of the ordinary emplastrum lyttæ or by the acetum cantharidis, or the liquor epispasticus, which are considered to act with more celerity. A large blister has sometimes appeared to be of service, but vesicants should not be applied at the outset of the attack. They are most advantageous when the initiatory symptoms are on the decline, when there is not such high arterial action, and when the surface has become cooler. I have seen them do harm when applied too early. The blistered part may afterwards be dressed with savin ointment, by which means a modified and beneficial amount of counter-irritation can be continued.

When the stomach is so irritable that scarcely anything can be retained, *hydrocyanic acid* in an aqueous mixture, with a little glycerine or mucilage added, is one of the best of remedies. *Effervescing draughts* with the bicarbonate of potash and citric acid are sometimes given, but the evolution of carbonic acid gas by distending the organ makes it contract upon itself, and the contents are again pumped up. There is another objection to their use; as tympanites always in greater or less degree obtains, the distention of the stomach pushes up the diaphragm still higher, and renders the respiration more difficult; and, again, the neutral salt which is formed, by acting as an aperient, is liable to increase the peristaltic action of the bowels, a result which should be most sedulously avoided. When the tympany is very considerable a *fœtid injection* consisting of two drachms of the tincture of assafœtida in half a pint or a pint of decoction of pearl-barley may be administered; or an ounce of the oil of turpentine, first being made into an emulsion with the yolk of egg and then mixed with the same quantity of barley decoction as before mentioned, can be injected. The oil of turpentine taken in doses of ten or fifteen drops in some emulsion or bland drink, or five or eight grains of the compound galbanum pill, every six or eight hours, are good measures for adoption. When such do not produce the desired effect, O'Beirne's long elastic tube may be introduced high up into the bowel and there allowed to remain, by which means incarcerated gases find a ready way of escape and much comfort is experienced. It is when this condition of tympanites subsists, and gives great distress after the inflammation has ceased, that such measures are useful. When we do not feel certain that the inflammatory action has subsided, and when vesication has not removed the cuticle, terebinthinate embrocations are likely to be of service.

Constipation is another circumstance which in these cases generally obtains. A right and rational consideration of this matter is of cardinal importance, because the very wrong notion is sometimes entertained that the bowels must be moved, and under this erroneous reasoning drastic purgatives have been given, producing, as they were said to do, much mischief. The physician should bear in mind that the constipation is not the cause but often the *effect* of the inflammation, and that the indicated mode of procedure is first to subdue the inflammatory action, when in due time restoration of function will follow. To allay and mitigate peristaltic action — in other words to give rest to the parts in a state of lesion — is to carry out the same principle observed in enjoining the disuse of a torn muscle, and in peremptorily excluding light in the treatment of an inflamed eye. If it is believed that there is great accumulation in the colon, an enema with olive oil and half an ounce of the spirits of turpentine in decoction of barley may be administered by means of the O'Beirne tube, and such may be repeated if deemed necessary; but there is benefit in frequently having recourse to this remedy in order to keep up gentle action of the intestines. To give purgatives by the mouth is often to set up or augment the irritation in the gastric mucous membrane, and by increasing the peristaltic action in the bowels to aggravate the disease. The contents of the intestines are often but soft and pasty matters, and then their presence can do no harm. There is a far greater liability to error in being too solicitous respecting the movement of the bowels than in leaving them to the efforts of nature.

Diaphoretic and *diuretic* medicines are to be used with the foregoing. The acetate liquor of ammonia, the ætherial spirits of nitre with camphor julep, form a good mixture, and tend to keep the skin and kidneys in the performance of their functions. Small quantities of strong beef-tea or farinaceous food are to be given at intervals. Smoler of Prague gives a little broth once or twice daily, and as little drink as possible while the activity of the disease continues. Urgent thirst may be allayed by pieces of ice being put into the mouth.

Such, then, is the line of treatment to be pursued in the *sthenic* or more flagrant forms of inflammation of the peritoneum, but they are not often met with, and constitute exceptions rather than the rule. It would be out of place here to enter upon that troubled question, the change of type in disease, but certain it is, whether from agencies operating from without, or from causes originating in the organism itself, that depletion in this disease is very rarely warrantable in the way in which I have described; nevertheless it would be wrong to pass into that extreme of inertness which has of late become but too prevalent, for, as I believe, moderate blood-letting in rightly selected cases is yet, despite the conflictions of controversy and the caprice of fashion, a valuable remedy.

As observed, by far the greater number of cases of Peritonitis presented to our notice are of the *asthenic* type—in that adynamic state of the system that will not bear lowering, and in which the general strength should be husbanded, not destroyed: for instance, in such examples as are consecutive upon or the sequels of some foregoing malady, when following the eruptive fevers, when metastatic of erysipelas, when the complication of albuminuria, when it occurs in perforation of the bowel in enteric fever, in the bursting of a mesenteric gland, in phthisis abdominis, in those occult blood changes which affect general nutrition, as in cancer, struma, and the climacteric period, or cirrhosis and cardiac disease, and in contamination of the fluids, as in pyæmia and puerperal Peritonitis. When we have to treat it as related to such conditions, our remedial measures must be resolved upon with great modification. *Opium* in the asthenic form is the chief agent, and Drs. Graves and Stokes were among the first physicians who gave this drug very largely. An impression decided and speedy must be made upon the nervous and sanguiferous systems, and in such lies our main hope of arresting the disease. It should be given in larger doses, and the effect kept up in full and apparent manner, but not to the induction of narcotism. Two or three grains may at first be prescribed, and a grain every four, three, or even two hours afterwards. Some in very urgent cases give half a grain, or even a grain, every hour. But in these perilous attacks of illness the patient should be frequently visited, and the physician should cautiously watch the effects of the remedy. Narcotism will be produced much sooner and with a far less dose in some persons than in others. If there be much sickness, laudanum injections should at short intervals be administered, instead of giving the drug by the mouth. In cases of great prostration and debility, quinine and camphor may be conjointed with the opium. In *perforation*, when the contents of the bowel are liable to be extruded into the serous cavity, and when lymph is thrown out, by which means the conservative attempts of nature are to seal up the orifice and mend the breach, to subdue and still the action of the part is everything. Motion implies the pouring out of the intestinal matters, the removal of the lymphic plug—in other words, a fatal issue. To paralyze the bowel for a time is the aim, in order that reparation may be favored. In these particular cases I would not give mercurials by the mouth. If they were to increase the flow of bile, and thus augment the peristaltic action, they would do incalculable harm. Inunction, as above recommended, might be used until the gums became slightly affected. It is far better to depend upon opium. In perforation there is sometimes very great tolerance of this drug. Murchison has known so large a quantity as sixty grains to be given in three days with impunity. In traumatic wounds, in the operation for hernia, and in paracentesis abdominis, the same kind of treatment should be followed. Fomentations, turpentine stoups, or a large poultice may at the same time be employed. Subsequently vesication may be ordered—and such repeated according to circumstances.

In that kind of Peritonitis complicated with Bright's disease, the primary complaint should be more regarded than the intercurrent affection. Salivation is to be carefully avoided; diaphoretics, warm cataplasms, rubefacients to the loins, warm baths, the hot-air bath, vesicants, and nutrients are then indicated. When the acute symptoms have subsided, the compound jalap powder and Dover's powder may be given. When the attack follows the exanthemata, is metastatic of erysipelas, or connected with pyæmia, mercury is inadmissible.

In *puerpral peritonitis* the treatment is often difficult and doubtful. and it should earnestly be borne in mind that it is frequently associated

14

with or consecutive upon an altered or vitiated condition of the blood If the power of the pulse warrant the lancet, bleeding, to be of benefit, should be done *early*. If deferred it is likely to do harm. The best authorities are emphatic on this point. Dr. Ferguson asserts that to be beneficial it must be employed within the first twenty-four hours, and that in the second stage of the disease it often produces a rapidly fatal result. Churchill is of opinion that when the remedy is admissible the time for its beneficial use is very limited, and he has seen no good from its employment after the first twenty-four hours. The first-named physician in doubtful cases gave ten grains of Dover's powder, and covered the abdomen with a linseed-meal poultice, which from its thickness would keep warm for four hours. At the expiration of that time, if the symptoms were not relieved, ten grains more of Dover's powder and another poultice were prescribed. If in other four hours from this second medication the malady did not yield, he had recourse to depletion. Sometimes when the pain is great and the pulse tolerably firm, two or three dozen leeches at once applied and followed by fomentations give good results. In the majority of cases, measures will be required which have previously been described as suitable to the asthenic type of this inflammation.

In the *Peritonitis of children* those general principles are to be aimed at which have already been given. It need scarcely, however, be more than mentioned here that these little patients always require their maladies to be managed with a gentle hand, and most especially in the use of depletion and opiates. These remedies with them are very uncertain in their effects, and sometimes produce a far greater impress upon the general powers than calculated upon by the practitioner. The age, the history of the case, and the cardinal signs will be our guide, and our measures should be modified according to the facts and exigencies of each particular instance. In the sthenic types, leeches, calomel, and if the age permit, carefully regulated doses of opium, linseed-meal poultices, terebinthinate epithems, warm baths, and injections are to be used. When the affection comes on as the sequel of one or other of the eruptive fevers, if we believe it to be traceable to some constitutional malady, some depravity of the organism, depletion and antiphlogistic means will be unwarrantable; then mercurial alteratives, small opiates, fomentations, warm baths, and counter-irritation will be the best measures. When the little patient tides over the more perilous days of active disease, and the case drifts onwards towards the more chronic condition, and when we find that there is effusion, counter-irritation and mild mercurial alteratives should be given, and during convalescence the iodide of potassium with decoction of sarsaparilla, the syrup of the iodide of iron, or quinine with the tincture of the perchloride of iron, often produce excellent effects. In the strumous diathesis cod-liver oil may be prescribed.

It has in this article been previously pointed out to the reader that Peritonitis not seldom occurs in a partial manner, and as a *complication* arising in the course of some foregoing disease, as when an antecedent malady, first instituted in some organ or organs covered by the peritoneum, is at length extended to it. For instance in hepatitis, when the convex surface is the seat of lesion it remains circumscribed; or the inflammation may be extended through to the pleura, and pleuro-pneumonia result, as in a case which I recently witnessed. It is then quite clear that our remedies should be addressed to the viscera involved, as well as to the serous membrane. In acute splenitis the turgor of that viscus should be relieved, or it would be vain to try to mitigate the peritoneal symptoms, which have their origin in the stretched, tense, irritated condition of the capsular coverings. In the liver affection we should as soon as possible bring to bear the influence of mercurials; but in diseases of the spleen, mercurials are most improper and would do harm. It is incontestable then that our diagnosis must be rightly formed, or our practice will be incorrect. In diarrhœa and dysentery, when associated with an inflamed peritoneum it is needful at once to control the excessive action of the bowels, and when such is subdued, the irritation extended to the serous membrane is likely to be subdued also. Opiate enemata, fomentations, the compound ipecacuan powder, and counter-irritants are the best measures. It has been remarked that the right iliac fossa is often the seat of pain, the disease being located near the cæcum, and it sometimes happens that the impaction of indurated fæces has much to do with setting up the inflammation. Large bland enemata, by unloading the great bowel, are in such cases of excellent service. When the sexual and urinary organs are first affected and Peritonitis becomes superadded, the primary disease should be held in view, and by its mitigation or removal the consecutive complaint will be benefited. From all, then, which has been said, it is obvious that in the treatment of every case the successful issue will greatly depend upon a clear and correct conception of the nature of the ailment, and a right interpretation of those symptoms which indicate the particular kind of morbid changes which obtain.

When the more acute stage has passed over, and those remedies suited to the earlier period of the attack have been employed, small doses of opium may still be given in combination with quinine or some of the bitter infusions. The various preparations of iron are of great value, and perhaps the tincture of the perchloride is the best. It is safest to defer as long as possible the use of aperients, and in preference the gentle action of the bowels should from time to time be promoted by bland enemata. When the active state of the affection has quite ended an occasional dose of gray powder with rhubarb and the bicarbonate of soda may be given. Terebinthinate and other stimulant embrocations can be applied to the abdomen when there is effusion, and a flannel bandage round the body, so as to ensure moderate and well-regulated pressure, is another mode of favoring absorption.

The *diet and regimen* during convalescence are of great importance. At the first soups and farinaceous food are to be allowed, and for some time solids should be interdicted. Arrow-root, tapioca, the Indian cornflour, with milk, are nourishing; and veal or chicken-broth with the crumb of bread may be given; and in the course of time beef-tea with toast, boiled chicken, and pounded meat may be taken. When stimulants are needed, sherry, weak brandy and water, claret, and bitter ale may be allowed. Flatulent vegetables and acescent fruits should for some time be discarded. An occasional warm bath to keep the skin in proper action is desirable. When the patient shall have so far recovered as to be able to travel, change of air will generally expedite his restoration to health.

TUBERCLE OF THE PERITONEUM.

By JOHN SYER BRISTOWE, M.D., F.R.C.P.

PATHOLOGY.—The deposition of tubercular matter in connection with the peritoneal membrane is of very common occurrence. For generally in cases of tubercular ulceration of the bowels, and certainly in all those cases in which the ulceration is extensive, gray granulations may be found in greater or less abundance studding those areas of serous surface which correspond to the areas of mucous ulceration. But tubercular formations of this kind seldom show any tendency to spread, and are rarely productive of appreciable mischief. They are for the most part, indeed, purely local phenomena.

There are other cases, however, far less common yet still not infrequent, in which the tendency to the deposition of tubercle is general throughout the serous membrane, and in which ulceration of the bowel is evidently not the starting-point of the peritoneal affection, and indeed is often altogether absent. To these, which were formerly known as mere varieties of chronic peritonitis, the name of Tubercular Peritonitis is now very often given. They are characterized not only by the comparative severity and extent of the peritoneal affection, but also by the fact that the symptoms of this affection are usually well-pronounced, and sometimes indeed are paramount.

Tubercular Peritonitis, like tuberculosis generally, may occur at any age, but is probably most common in early life. Out of 48 cases extracted from the records, for a limited period, of St. Thomas's Hospital, 3 were under ten, 14 between ten and twenty, 13 between twenty and thirty, 9 between thirty and forty, 7 between forty and fifty, and 2 between fifty and sixty. But in correction of these figures it must be recollected that children under ten are admitted in small proportion into general hospitals. Out of the same number of cases, 26 were males, 22 females; but 222 tubercular males were admitted to 127 tubercular females, and proportionately to this number tubercular peritonitis was more frequent in the female than in the male, in the ratio of very nearly three to two. In two cases only was the tubercular deposit limited exclusively to the peritoneum. In all the others—namely, in 46 cases—there were tubercular deposits in other organs, and generally in several other organs. In 42 there was tubercle in the lungs; in 25, in the intestines; in 25, in the pleuræ; in 20, in the spleen; in 14, in the bronchial glands; in 11, in the kidneys; in 10, in the mesenteric glands; in 9, in the liver; in 8, in the brain; in 4, in the uterus and Fallopian tubes; and in 1, in the pericardium. But taking into consideration the relative frequency with which the several organs just enumerated are the seats of tubercle, a very different numerical relation than that just given becomes apparent between tuberculosis in them severally and tuberculosis of the peritoneum. Thus tubercular disease of the peritoneum was present in (to disregard fractions) 74 per cent. of cases of tubercle of the pleura, in 53 per cent. of cases of tubercle of the spleen, in 46 per cent. of cases of tubercle of the kidneys, in 44 per cent. of cases of tubercle of the brain and of the uterus and Fallopian tubes respectively, in 39 per cent. of cases of tubercle of the liver, in 37 per cent. of cases of tubercle of the bronchial glands, in 33 per cent. of cases of tubercle of the pericardium, in 29 per cent. of cases of tubercle of the mesenteric glands, and in 12 per cent. only of cases of tubercle of the lungs and of tubercle of the intestines severally. It may be worth while to add, that out of the 46 cases in which there was tubercular deposit in other organs besides the peritoneum, the most serious lesion was in 12 the tubercular disease of the peritoneum; in 15, that in the lungs; in 8, that in the brain; in 3, that in the pleura; and in 1, that in the intestines.

Peritoneal tubercles present much the same characters as tubercles occurring in other parts. They are sometimes miliary, or in the form of minute roundish spots, varying from mere points up to the size of a poppy-seed, and having an opaque white, or grayish or yellowish aspect. Sometimes they form rounded or lobulated masses, from the size of a tare up to that of a hazel-nut, presenting for the most part an opaque buff color, studded often with black points or patches, and exhibiting a cheesy aspect and consistence which are modified by the more or less abundance of fibroid material which invests and permeates them. Sometimes again, but much more rarely, there are found, lying between organs which are adherent, tubercular laminæ of considerable thickness and extent. Peritoneal tubercles exist rarely, if ever, independently of the effusion of lymph, and indeed rarely, if ever, are formed otherwise than in the substance of such adhesions, although they may subsequently in the progress of enlargement involve not only the peritoneum itself, but the tissues which are subjacent to the peritoneum. There is probably no essential distinction between the miliary form of tuberculosis and that in which the tubercles form masses of larger size: the former, however, are most frequently found in cases of acute progress, the latter in cases which are chronic; the former, moreover, are generally discovered thickly-set and innumerable, the latter in comparatively small numbers. In cases of miliary tuberculosis indeed, the peritoneal surface is mostly found covered with a layer, of various thickness, of grayish, transparent, adherent and toughish lymph, which not only invests the abdominal organs, but renders them more or less adherent to one another. And in the substance of this lymph the tubercles are disseminated as opaque grains, which may be separated with the lymph from the subjacent peritoneal surface. This condition may be general, or it may be limited to certain regions, and not infrequently when thus limited the parts affected are studded with filaments of lymph, in which miliary tubercles may be recognized. In the other form of the disease, the peritoneal surface is covered with lymph which has assumed the form of connective tissue, and the adhesions between organs are formed of tough fibrous bands. And it is in this tissue, and especially among these bands (sometimes forming the centre of a kind of knot, sometimes forming flattened masses between closely united surfaces), that the large masses of tubercle are for the most part found. It is this form of tubercle which occasionally invades the intestinal walls, and leads to perforation of the bowel from without. In association with the deposition of peritoneal tubercle, the various accompaniments and sequelæ of common inflammation manifest themselves generally in a greater or less degree. Thus, there is often patchy and streaky redness, often fibrinous effusion which is not visibly tubercular, and often effusion of serum; sometimes there is suppuration, and sometimes again hæmorrhage into the peritoneum. The most important of these, from its frequency, is undoubtedly the effusion of serum. Indeed tubercular disease of the peritoneum is a common cause of ascites. It is probable that most cases in which tubercle exists on the peritoneal surface prove fatal sooner or later, either from the direct effects of the peritoneal disease or from the effects of tuberculosis in other organs. Yet there can be no reasonable doubt that recovery sometimes takes place. For not only does our knowledge of the progress of tubercle in the lungs justify us in this inference, but we not infrequently meet with cases of recovery from symptoms which we have the strongest reasons to regard as dependent on tubercular peritonitis, and still more, we occasionally detect in the abdomens of persons dead of other diseases signs of old peritonitis, together with the presence of earthy nodules such as result from the drying up of tubercle.

SYMPTOMS.—The symptoms which attend the progress of peritoneal tuberculosis present much variety, and are often vague and indefinite. Often, indeed, and not only in those cases in which the peritoneal affection is slight, or in those in which it is as it were overshadowed by the preponderance of disease in other parts, but in those cases even in which it is the predominant or sole affection, they fail to indicate clearly the peritoneum as the seat of any disease. Further, they are so generally complicated with the symptoms which are due to co-existing tubercular disease in other organs, especially in the lungs, pleuræ, and intestines, that it is impossible altogether to dissociate them from these latter.

Most cases, however, of tubercular peritonitis, in which there are obvious indications of abdominal disease, may be arranged, somewhat roughly perhaps, in two classes: the first, the acute class, in which the symptoms bear a considerable resemblance to those of enteric or of so-called "remittent" fever; the second, the chronic class, in which the symptoms correspond for the most part with those of " chronic peritonitis." In the acute form of the disease, the patient sometimes in the midst of perfect health, more often however after some indefinite period of languor and loss of flesh and strength, begins to manifest febrile symptoms attended with remissions and indicated by heat and dryness of surface with quickened pulse, pains in the limbs and loins and head, diminution of the secretions, and perhaps drowsiness. At the same time the abdomen probably becomes somewhat hard and tumid and tender, and the patient complains of more or less pain in it. Generally also there is some disturbance of the digestive functions, dryness or furring of the tongue, thirst, loss of appetite, and nausea or sickness, with probably constipation or diarrhœa, or an alternation of these conditions. And with no material change in these symptoms, perhaps, beyond that which is due to gradually increasing debility and emaciation and the gradual supervention of what are ordinarily known as "typhoid symptoms," the patient gradually sinks, and at the end of a few weeks dies. The distinctions between acute abdominal tuberculosis and enteric fever consist, as regards the former disease, partly in the absence of rash, the less constant disturbance of the bowels, the non-limitation of tenderness to the cæcal region, and the less definite duration of the disease, and partly in the occasional presence of characteristic complications, among which may be enumerated tubercle in the brain, pulmonary phthisis, renal disease with albuminuria, and the accumulation of ascitic fluid. It may be remarked, however, that even in spite of care the cerebral symptoms arising from tubercle in the brain may be mistaken in some cases for the delirium of enteric fever, and the symptoms of pulmonary tuberculosis may pass for those of the pulmonary affections which so commonly ensue in that fever; and further, that the liability in both cases to intestinal perforation and acute peritonitic symptoms furnishes an element of serious difficulty in reference to diagnosis. In the chronic variety of peritoneal tuberculosis, the disease sometimes commences with more or less typical symptoms of acute peritonitis, sometimes creeps on with the utmost insidiousness; but in both cases (in the one after the disease has become fully established, in the other after the acute initial symptoms have subsided) the symptoms gradually become more or less identical with those which have been described elsewhere as indicative of chronic peritonitis: symptoms which, with many variations and remissions and exacerbations, may continue for a month or longer, and upon which in most cases sooner or later ascites supervenes. It must not be forgotten that in the chronic, as well as in the acute affection, deposition of tubercles in other organs is apt to take place, and that in its course the presence of tubercles in the brain, lungs, bowels, or elsewhere, may produce symptoms which may lead us or mislead us in our diagnosis; and that in this case, even more than in the other, there is liability to tubercular perforation of the bowel, and to lardaceous or other degenerative diseases of important organs, especially of the liver and the kidneys.

As examples of some of the many anomalous cases which do not by their symptoms fall very obviously under either of the above categories, I may here briefly quote two cases. A girl about twenty had been ailing for some twelve or fifteen months. She had been getting weak and thin, and had been suffering from attacks of severe sickness, coming on with some regularity every three or four days. The sickness was remarkable from the facts that during the three or four hours for which it lasted she would bring up as much as a couple of wash-hand-basinfuls of nearly clear fluid, that it was apparently independent of the ingestion of food, and that between whiles she had no symptoms of indigestion and had a good appetite. There was, further, no affection of the bowels, and no distinct abdominal enlargement or tenderness. These symptoms continued while she was under my care; but shortly after she came under my care, and then for the first time, a cough came on, consolidation was discovered under the left clavicle, and from that time pulmonary consumption made rapid progress. Her death, which was mainly caused by the pulmonary disease, occurred about three months after I first saw her; and at the post-mortem examination there was found, in addition to extensive tubercular disease of the lungs, very extensive peritoneal tuberculosis. The stomach and bowels were healthy.—A young gentleman of two or three and twenty, who was at the time resident at Port Natal, became without any apparent cause subject to attacks of intense colic, in which he was compelled by the severity of the pain to throw himself down and writhe. He came over to England in consequence of the persistence of this affection. The attacks of pain still continued, coming on sometimes two or three times a day; but there was also some irregularity of the bowels. His illness lasted for about a couple of years, and he died then from emaciation and exhaustion. There was more or less general tuberculosis discovered after death; but the chief deposit was in connection with the peritoneum. Occasionally the chief symptoms due to the presence of peritoneal tubercle are great obstinacy of the bowels, with gradually increasing emaciation and debility; and occasionally there is complete and insuperable obstruction. In some cases, ascites is the earliest prominent symptom, and it may continue the most prominent symptom, and then prove (as ascites from other causes often proves) the chief agent in causing death.

CARCINOMA OF THE PERITONEUM.

By John Syer Bristowe, M.D., F.R.C.P.

PATHOLOGY.—Carcinoma of the Peritoneum, using the term in its widest sense, is not infrequently met with. Taking for comparison the same period which furnished, from the medical wards of St. Thomas's Hospital, 349 cases of tuberculosis of which 49 presented peritoneal complications, there were 99 cases of cancer, in 22 of which the peritoneal membrane was affected. From these figures it would appear that while cancer of the peritoneum is less than half as common as tuberculosis of that membrane, it is considerably more common in reference to all cases of cancer than tubercle of the peritoneum is to all cases of tuberculosis.

There is probably no great difference, in the liability of the two sexes to this disease; but there is no doubt, I think, that it is relatively less frequent in early life than tuberculosis. Of the 22 cases alluded to above, none occurred under twenty, 3 occurred between twenty and thirty, 4 between thirty and forty, 5 between forty and fifty, 5 between fifty and sixty, and 5 between sixty and seventy. In 2 cases the disease was apparently limited to the peritoneum, or had at most invaded the surface of organs invested with the peritoneum. It was associated in 11 cases with cancer of the stomach, in 10 with cancer of the liver, in 9 with cancer of the pleuræ, in 7 with cancer of the lungs, in 6 with cancer of the mesenteric glands, and in 3 severally with cancer of the bowels, kidneys, and ovaries. More than half the cases, however, of cancer of the bowels were combined with peritoneal cancer; rather less than half the cases of cancer of the pleuræ and stomach respectively were associated with it; and about a fourth of all cases of cancer of the liver, mesenteric glands, kidneys, ovaries, and lungs respectively presented the same complication. In 7 cases the peritoneal cancer was the predominant disease, in 10 cancer of the stomach, in 1 cancer of the liver, and in 1 cancer of the mesenteric glands. It may be added here, that in speaking of peritoneal cancer, those cases have been excluded in which that portion of peritoneum covering a cancerous organ has alone presented indications of cancerous growth.

Carcinoma of the Peritoneum presents most of the varieties which carcinoma presents in other parts of the body; namely, scirrhus, encephaloid (with its sub-variety melanotic cancer), and colloid. Scirrhus always commences in the form of flat, round, lenticular, hard, white spots, measuring perhaps on the average a line in diameter, which occupy the substance of the serous membrane, and though distinctly projecting from the surface, yet rather tend to invade and involve the sub-serous tissue. These are in the first instance scattered thinly or irregularly, but soon become aggregated in parts or generally, and then coalesce so as to form patches

of various extent. The patches thus formed may be perfectly smooth on
the surface, or may still present there traces of the mode in which they
were originally formed; they rarely, however, form outgrowths, and pretty
rarely invade subjacent organs; rarely, too, over the general peritoneal
surface do they become more than a line or two thick, except when they
involve duplicatures or processes of peritoneum. The latter involvement
is indeed somewhat characteristic of the disease. The appendices epiplöicæ
become converted into small hard masses, in which the cancerous deposit
and the fat and other normal tissues become intermixed; the mesenteric
and other like duplicatures become often similarly affected; and the great
omentum, from the same cause, becomes contracted into a thick band,
stretching transversely across the abdomen in the course of the transverse
colon.   Scirrhous cancer, in fact, as has long been recognized, tends rather
to cause contraction of parts than outgrowths: and for this same reason
has a special tendency not only to cause the contractions of loose tissues
already adverted to, but to lead to obstruction of tubular organs, especially
of the stomach, intestines and larger bile-ducts. Encephaloid also in its early
stage affects the substance of the peritoneum, and forms discrete nodular
outgrowths, which are small and rounded, and differ from those of scirrhus
not only in their greater softness, but also in their greater prominence. These
are often indeed hemispherical, or even spherical or pyriform and pedunculated. In its further progress encephaloid presents great varieties. In some
cases it seems, like scirrhus, to invade more particularly the substance of
the peritoneal folds, and to involve also subjacent organs; and under such
circumstances we find sometimes the mesentery converted into a thick,
plicated, cancerous mass, with the cancerous growth extending from the
mesenteric attachment over the surface of the intestines; or we find the
greater or lesser omentum or the sub-peritoneal tissue of other regions affected in like manner, and forming a more or less distinct tumor. In
other cases it tends rather to form outgrowths which are sometimes small
and clustered, sometimes more or less distinct from one another and
rounded and massive. In the former instance the whole peritoneal surface
may be found beset with small lobulated or bunch-of-currant-like excrescences, and the great omentum may be converted into a huge loose
mass of such bodies. In the latter instance the tumors, though more or
less abundant, are isolated, and while many probably are small, others
form rounded solid masses which may attain the size of a child's head.
So far as I know, melanotic cancer always manifests itself in this latter
condition. Colloid disease in its early stage appears for the most part in
the form of groups of vesicles which vary in fineness and have a close
primâ facie resemblance to patches of eczema or herpes, or (if the fibroid
element be abundant) in the form of slightly granular or delicately-reticu-
lated patches. Later on, the vesicle-like bodies are often as large as a
millet-seed or tare. The patches often become more or less elevated above
the level of the surrounding surface, and spread sometimes in tortuous
and anastomosing lines as though taking the course of the lymphatic vessels, sometimes by forming scattered, isolated, somewhat pedunculated
growths. This disease, like scirrhus and encephaloid, tends in various
degrees both to involve subjacent organs and to diffuse itself over the
peritoneal surface. It always involves the sub-peritoneal tissue, which
may attain in consequence very considerable thickness; and it extends
thence most frequently to the muscular and mucous coats of the stomach
and intestines, less frequently to the substance of the mesenteric glands,
pancreas, liver, spleen, or other viscera. In the most extreme cases of

the disease, nearly the whole of the peritoneum is affected; this membrane is then irregularly thickened, with lumpy excrescences here and there; the various duplicatures become especially hypertrophied; and the great omentum is sometimes converted into a huge lobulated mass, or is contracted, as it generally is in scirrhus, into a thick irregular transverse band. In all these cases the adventitious growth retains its original more or less distinctly vesicular if not gelatinous character; and generally, sooner or latter, from erosion of its surface, the glairy fluid contained in its substance is discharged in some abundance into the cavity of the abdomen.

Other varieties of cancer, such for example as osteoid cancer, are probably always secondary, and are of such extremely rare occurrence as to be of no practical importance.

All forms of abdominal cancer are liable in a greater or less degree to various complications. Among which may be enumerated: peritoneal inflammation, with the effusion of lymph or pus, or the escape of blood; ascites; obstructions of stomach or bowels; involvement of the viscera, such as the liver or kidneys, or their excretory ducts; and perforations of the stomach and intestines or other hollow organs.

SYMPTOMS.—The symptoms of peritoneal cancer are necessarily very various and often quite as easy to be misunderstood as those of peritoneal tubercle. Febrile symptoms, varying in intensity and liable to remissions, gradually increasing debility and emaciation, more or less uneasiness or tenderness or pain in the abdomen, with hardness and enlargement of the same part, disturbance of the functions of the alimentary canal indicated by dry and glazed or coated tongue, thirst, loss of appetite, with perhaps nausea and sickness, and by constipation or diarrhœa or alternations of both, are symptoms which are common alike to cancer and to tubercle and to mere chronic inflammation of the peritoneum. It is important, however, to bear in mind that obstinate constipation is a very frequent accompaniment of this disease, and that much more frequently than in either turberculosis or inflammation, death results from complete obstruction; also, that in a very large proportion of cases the stomach is involved in a greater or less degree, and that consequently the usual symptoms of stomach-cancer are very liable to be associated with those of the peritoneal affection; further, that in nearly half the cases there is cancer of the liver, not infrequently involving that organ through the gastro-hepatic omentum and Glisson's capsule, and that therefore obstruction of the bile-ducts and jaundice are of common occurrence; and lastly, that in the female there is frequent co-existence of ovarian and peritoneal cancer. The most important points, however, to which we must look for the formation of a correct diagnosis are, first, the presence of a growing tumor or tumors in the abdomen, and, second, the presence of similar disease in other parts. It need scarcely be said that cancerous tumors present all varieties of character; that they may occur in any region of the abdomen; that they may be movable or fixed; that they may vary widely in size and shape; that they may be hard and resisting, or soft and almost yielding a sense of fluctuation; and that, especially when they are developed in the neighborhood of the cœliac axis and superior mesenteric artery, they may pulsate as violently as many aneurisms do; and that hence notwithstanding the important aid which their presence furnishes, they may be, and are not infrequently, confounded, at some stage at least of their progress, with circumscribed abscesses, or hydatid tumors, or floating kidneys, or even aneurisms. But in some cases where, although the cancerous disease is very extensive, the individual tumors are small, the presence of the peritoneal outgrowths may fail of detection, even when very care-ful examination has been made; and necessarily this difficulty of detection is always greatly increased when ascitic fluid is present. It is worth while to draw attention to the fact, that not infrequently when no other signs of tumor are distinguishable, the presence of the thickened and con-tracted great omentum, which has been shown to be common in scirrhus and in colloid disease, may be recognized as a more or less irregular trans-verse bar extending horizontally from under the margins of the left ribs across the upper part of the umbilical region to the neighborhood of the umbilicus, and that this furnishes a valuable diagnostic sign.

It is impossible to lay down any rules with regard to the detection of concurrent cancerous disease in other organs; but it is obvious that in all cases in which there is any ground to suspect that a patient may be suffer-ing from internal cancer, a careful investigation of all superficial and other easily accessible parts should be made; for not infrequently there may be found associated with the internal cancer, coming on before it, or appearing at a later period, cancerous nodules in the subcutaneous cellu-lar tissue, cancerous growths of periosteum, or bone, or cancer affecting the uterus, mamma, or testis. Nor must it be forgotten that cancer of the pleuræ, lungs, and mediastinum, cancer of the brain, and cancer of the kidneys, are all with different degrees of frequency apt to be associated with cancer of the peritoneum.

Treatment of Abdominal Tubercle and Carcinoma.—There are stages in many varieties of the diseases coming under the above heads, when, as has been shown, they may be readily mistaken for other affections of a less grave character than themselves; and when therefore it may be judicious to adopt the treatment, whatever it may be, which may seem most suitable for the more curable malady. But, assuming the fact of the presence of tubercle or of cancer to be known, the principles of treatment become ex-ceedingly simple: they are, to relieve pain and discomfort by ministering to those symptoms which most distress the patient, and to support his strength by the judicious exhibition of food and stimulus, and by the use of medicines having a similar tendency. Abdominal pains may need to be relieved by the application of counter-irritants, or fomentations, or even leeches. Sleepless weariness and pain may require to be overcome by the use of opiates or other forms of sedative or narcotic medicines; and indeed, in the progress of cancer especially, these remedies are often the only ones that can be employed, and may have to be given constantly and in large doses. Nausea, sickness, diarrhœa, obstruction of the bowels, will each in various cases call for treatment, but nothing special need be said in reference to them. That tonics, food, and stimulants, of such kind and in such quantities and at such intervals as the condition of the patient renders admissible should be persisted in is obvious, not only because the maintenance of life up to the extreme limits which the progress of the diseases admits of depends thereon — and it is our recognized duty as physicians to sustain life even when it is a hopeless burden—but because (to say nothing of the chance there may be of our diagnosis being in some cases erroneous) there may be, at least in the case of tubercular disease, a prospect, however remote, of ultimate recovery.

AFFECTIONS OF THE ABDOMINAL LYMPHATIC GLANDS.

By John Syer Bristowe, M.D., F.R.C.P.

The lymphatic and lacteal glands of the abdomen are frequently the seat of disease; sometimes they become inflamed, sometimes hypertrophied, sometimes tubercular, and sometimes the seat of the various forms of cancerous growth and of degenerative changes.

In inflammation they become enlarged, congested, softened, and tender, and sometimes undergo suppuration, and may then discharge their contents by various routes, and even by rupture into the peritoneum. When the inflammation subsides they may according to circumstances recover their healthy state, or remain enlarged, or become atrophied and indurated. The symptoms indicative of their inflammation are more or less pain and tenderness in the situation of the affected glands, with perhaps hardness or distinct tumor, and more or less violent inflammatory fever. Inflammation of the abdominal glands is probably of very common occurrence as secondary to inflammation or ulceration of the various organs with which they are in connection, but we are chiefly acquainted with inflammation of the mesenteric glands in enteric fever, and in dysentery, and of the lumbar glands and those about the brim of the pelvis in connection with inflammatory affections of the genito-urinary organs.

Hypertrophy of the glands is not very easy to separate from tubercular disease of the glands on the one hand, and from some forms of malignant disease on the other. It is indicated by a more or less gradual increase in their size, attended with a more or less fleshy consistence, and a color varying between a dull white or buff, and a reddish fleshy hue. It is an affection rarely limited to the glands of a particular part; and generally, therefore, when the abdominal glands are hypertrophied, the lymphatics of other parts of the body are hypertrophied also. The symptoms which attend this affection are rarely connected specially with the abdomen; excepting in so far as there may be a tumor there, and more or less impairment of nutrition; they are for the most part those of gradually increasing anæmia, and a form of cachexia, in which sometimes there is a remarkable increase of white corpuscles in the blood (Leucocythæmia).

Tubercular deposits, in the mesenteric glands especially, are not uncommonly associated with similar deposits in the peritoneum and intestines; and they generally form well-defined cheesy lumps embedded in enlarged and more or less congested gland substance. Not very infrequently such deposits take place in glands which have previously undergone hypertrophy, and to such an extent sometimes that whole glands become caseous. Tubercular glands sometimes soften or suppurate and form vomicæ; and very frequently indeed dry up and contract and become converted into inert cretaceous masses. This condition of glands is probably attended with no symptoms distinguishable from the symptoms due to the associated tubercular affection of other abdominal organs which is generally present.

Cancerous disease of the various abdominal glands is common in all its varieties. It is sometimes primary (in which case it is probably generally if not always some variety of what Virchow terms lymphoma). It is more frequently secondary to cancer of other parts; and then, for the most part, the glands chiefly affected are those which are in relation with the organ primarily affected. Thus, in cancer of the testis the lumbar glands become cancerous; in cancer affecting the remaining genito-urinary organs, and other organs situated in the pelvis, the glands which become specially implicated are those in the pelvis, and about its brim; in cancer of the bowels, the mesenteric glands chiefly suffer; and in cancer of the stomach, kidneys, and neighboring parts, the retro-peritoneal glands of the upper part of the abdomen. Cancerous glandular tumors sometimes attain an enormous size; and it is not infrequently by their growth and disintegration that perforation or obstruction of viscera, and other serious complications, which have been elsewhere sufficiently described, are produced. It is difficult, and would be useless, to discuss the symptoms and effects of such tumors apart from those of cancer of the peritoneum and other abdominal organs, which have been already fully considered.

In addition to the degenerations which follow upon inflammation, and upon the deposition of tubercle, it may be stated that in extreme cases of lardaceous disease, the abdominal lacteal and lymphatic glands may share with other parts in this form of degeneration.

ASCITES.

By John Syer Bristowe, M.D., F.R.C.P.

Pathology.—The accumulation of fluid of a more or less serous character within the peritoneal cavity is called "Ascites," or "Abdominal
Dropsy." It is an accompaniment or sequela of numerous different forms
of disease; but depends immediately on some condition which modifies the
action of the capillary vessels, and in some cases perhaps of the lymphatics, of the peritoneal membrane. This condition may be, in the first
place, some morbid process going on in the peritoneal tissue, and affecting directly its minute vessels; or, in the second place, some impediment
to the return of blood from them existing in the course of the portal system; or, in the last place, some impediment to the return of blood from
them connected with some disease affecting generally the movement of
blood in the systemic veins. Among the first of these classes may be
included peritonitis, peritoneal tuberculosis, and peritoneal cancer; among
the second, tumors or other growths obstructing the trunk or main
branches of the vena portæ, chronic congestion and induration of the liver,
lardaceous disease of that organ, and especially cirrhosis; and among the
last, heart disease, Bright's disease, some affections of the lungs, and perhaps some forms of anæmia.

(1) Acute peritonitis, like acute inflammation of other serous membranes, is doubtless attended in most cases with more or less effusion of
serum; but the effusion is rarely abundant and rarely amounts to what
would be recognized during life as Ascites. Not very infrequently, however, when the acute peritonitic symptoms have subsided, and the patient
appears to be convalescent or even well, abdominal dropsy slowly supervenes. Ascites is especially apt to occur in women in whom the peritoneal inflammation has been connected with some inflammatory condition of the pelvic organs. It is frequently associated with the growth of
cystic ovarian tumors; and is then in some cases due either to the occasional rupture of small superficial cysts, or to the establishment of more
extensive communications between the cavities of the ovary and that of
the peritoneum, and the discharge of fluid from the thus exposed secreting
surfaces into the abdominal cavity. In all these cases the peritonitis
assumes a sub-acute or chronic character. Tubercular deposits in connection with the peritoneal surface are another fruitful cause of Ascites. In
12 out of the 48 cases of tubercular peritonitis analyzed on a former
page, this condition was present, and several of them had been tapped.
Abdominal cancer, again, is frequently attended with dropsical effusion.
Of the 22 cases of peritoneal cancer previously considered, 11 had Ascites
in a greater or less degree; and it may be added, that dropsy not infrequently attends cancerous disease of the ovaries and other pelvic organs,

15

and of the mesenteric or retro-peritoneal glands. In what degree Ascites, dependent on disease of the peritoneal membrane, may be due severally or collectively to direct involvement of the capillaries and minute veins of that membrane, to obstruction of the lymphatic orifices which seem now proved to exist there, or to increased functional activity on the part of the epithelial cells, is not very easy, perhaps, to decide; but there is probably little doubt that in some cases in which there is infiltration and contraction of the peritoneal folds, especially of the mesentery, the larger veins contained within them become, as Oppolzer suggests, obstructed, and that the Ascites is produced or augmented by this obstruction.

(2) Impediment to the passage of blood along the portal vessels, with consequent Ascites, may be caused by various morbid conditions; occasionally by the pressure on the vena portæ of an aneurismal, hydatid, or cancerous tumor, originating externally to the liver; more frequently by the pressure of cancerous, "knotty," syphilitic or hydatid tumors developed in the hepatic substance, and especially by cancerous and fibroid growths occupying the lesser omentum, and extending thence into the liver along the capsule of Glisson; but most commonly by some general hepatic disease which involves the hepatic capillaries and the minute veins which open into and emerge from them. Of the diseases last referred to, cirrhosis is the most frequent and the most important. Cirrhosis, however, though doubtless tending in all cases ultimately to cause Ascites, is sometimes fatal by hæmatemesis before any dropsical effusion has taken place, and is not infrequently found to be present, unsuspected, in death from other visceral diseases. Out of forty-six cases in which cirrhosis was discovered *post mortem*, in twenty only was there more or less accumulation of ascitic fluid. The presence of a fibroid capsule, surrounding the liver, compressing it, and squeezing it into a comparatively small rounded mass, produces the same effect. This formation, which is probably of inflammatory origin, is sometimes associated with cirrhosis, or other morbid states of the liver, but is sometimes present when the liver seems otherwise perfectly healthy, and where it is the sole visible pathological phenomenon associated with Ascites. There is no doubt that lardaceous disease of the liver also sometimes leads to abdominal effusion, and not improbably an extreme state of fatty deposition may have the same result; but in both of these cases the hepatic affection is almost always associated with still more serious disease in other organs, which is itself capable of causing dropsy, so that the influence of the liver in its causation is rendered somewhat difficult of identification. Similarly, it is quite certain that chronic induration and congestion of the liver, and especially that condition of the organ to which the name " nutmeg liver " is applied, are frequently instrumental in the production of Ascites, although they are themselves always secondary to dropsy-producing diseases, such as kidney disease, heart disease, chronic bronchitis and chronic phthisis.

(3) All the diseases which have just been enumerated, viz. chronic bronchitis and phthisis, heart diseases, and certain forms of kidney disease, which cause anasarca, cause naturally, as a part of that anasarca, effusion of serum into the abdominal cavity: but in most cases the abdominal effusion is proportional only to the effusion in other parts, and fails to be recognized as Ascites. In some cases, however the dropsical accumulation in the abdomen becomes excessive, while that elsewhere undergoes but little increase. When this happens, it is usually in connection with, and then probably immediately dependent on, some abdominal complica-

tion of the primary disease, especially a congested or indurated, or nutmeg, or even a cirrhosed condition of the liver, or chronic inflammation of the general peritoneal surface, or of that of the liver. But sometimes, even where the ascitic fluid has been so abundant as to need removal by operation, no trace of disease in any of the abdominal tissues or viscera can be discovered. There can be little doubt that in some forms of cachexia and anæmia, in which without there being any apparent visceral disease anasarca takes place, Ascites also occasionally ensues. Yet it may be remarked, that as cases of this kind usually get well, it must generally remain a matter of uncertainty as to whether or not there may have been some slight inflammatory affection of the peritoneum, or some other evanescent local morbid condition on which the Ascites may have depended.

It may be added here, that in a very large proportion of cases of Ascites, several or even many organs are diseased at the same time, so that it becomes difficult or impossible to determine upon what exactly the ascitic accumulation depends. Thus fibroid and lardaceous and other degenerations often affect simultaneously many organs, so that together with the liver we often find the kidneys, the spleen, the lungs, the heart, the blood-vessels, diseased in various degrees. Besides which, in all such cases there is a great tendency to inflammatory implication of the peritoneum as well as of other serous membranes; and tuberculosis is often present. This simultaneous affection of many different organs and tissues is specially common among those who have passed a life of debauchery, among those who have labored under the syphilitic cachexia, and among those who have suffered long from bone-disease, from protracted suppuration, or from chronic tuberculosis.

The amount of fluid present in Ascites may vary from a few pints up to four or five gallons, and indeed much larger quantities are recorded as having been met with. The fluid itself is for the most part slightly viscid, transparent, of a yellowish or greenish tinge, alkaline and containing both albumen and fibrine (or fibrinogen). It may, however, under different circumstances, become very viscid, opaline, or opaque from inflammatory products, or it may contain blood.

It would be tedious and, it is feared, useless to go at any length into the statistics of Ascites; for in the first place Ascites is an incident only of many different forms of disease, the statistics of which, with those of their particular relations to abdominal dropsy, are all elsewhere sufficiently discussed; and, in the second place, to bring together the statistics of Ascites in the gross, would be to combine a number of heterogeneous figures the manipulation of which could for the most part only lead to useless or fallacious results. There are a few facts, however, which the statistics of a general hospital have supplied me with, which it may be worth while to state. According to these statistics, there is little difference between males and females as regards their respective degrees of liability to Ascites, although undoubtedly hepatic dropsy is far more common in men than in women; Ascites is most frequent in the decades from thirty to forty and from forty to fifty, next in those from twenty to thirty and from fifty to sixty; but it is not uncommon, both above the latter age and in young children; it occurs with about equal frequency as the result of hepatic disease, heart disease, and kidney disease (in the latter two cases, however, generally combined with a congested or nutmeg or contracted condition of liver); it is from about one-half to one-third as common as a consequence of peritoneal cancer, peritoneal tubercle, bronchitis and phthisis severally; and, again, occurs in association with lardaceous

disease of organs and ovarian cystic tumors respectively about half as frequently as in connection with each of the immediately foregoing diseases.

The prospect of the duration of Ascites, and of eventual recovery or
death, necessarily depends almost entirely upon the nature of the disease
on which the dropsy depends. Now, most of the diseases causing abdominal dropsy are from their nature lethal, and generally, therefore, Ascites must be regarded as a symptom terminable only with death. Yet
even in some of these cases it is of very protracted duration, and relief
may be afforded several times by tapping before the arrival of the fatal
issue. But in some cases, and even when the disease causing it is usually
a progressive disease, in chronic peritonitis, in cirrhosis, in the encapsuled
state of liver, and probably also in tubercular peritonitis, the dropsy may
be sometimes arrested in its progress, or even, temporarily at least, recovered from. In some cases indeed, both in children and in adults, recovery from Ascites (the cause of which thus necessarily remains more or
less obscure) is permanent.

SYMPTOMS.—The symptoms due to Ascites alone are very simple and
very characteristic of the affection. The accumulation of fluid within the
abdominal cavity causes the abdomen to enlarge and become tense, and
then sooner or later compresses and obstructs the intra-abdominal veins,
especially those connected with the lower extremities, impedes the movements of the diaphragm, inducing difficulty of breathing, and interferes
more or less injuriously with the healthy action of the abdominal viscera.
It modifies also the patient's gait, making him walk like a pregnant
woman, with his legs wide apart, and his head and shoulders thrown back.

The presence of fluid in the peritoneal cavity is generally easy of detection. The abdomen becomes large, uniformly rounded, but with a
tendency to spread or bulge in the flanks as the patient lies on his back,
tense and more or less smooth and shining, often presenting distended
superficial veins and the linear lacerations of the deeper tissues of the
skin which are so common in pregnancy. The stomach and intestines
being lighter than the fluid, tend to float on its upper surface; and hence
generally the highest part of the abdomen according to the patient's position, is resonant, while the more dependent parts are dull, the line of
demarcation between them being for the most part well-defined and horizontal: hence, too, as the patient changes his position, the fluid and the
floating bowels, and necessarily therefore the areas of resonance and dulness, change their positions relatively to the abdominal parietes. It may
be added that the liver, which is generally if not always of higher specific
gravity than dropsical fluid, retreats sometimes distinctly, as the patient lies on his back, from the anterior surface of the abdomen, a stratum
of fluid with sometimes a loop of floating bowel occupying the interval.
The presence of fluid is further and very importantly indicated by the
peculiar thrill which is experienced by the hand laid flat on the abdomen
when a ripple or wave is produced in the ascitic fluid by a slight tap or
fillip applied to some other part of the abdominal surface. These signs,
however, are not always all present, or at least easy to recognize: and not
infrequently tumors and other forms of disease simulate or mask abdominal dropsy. Thus when the ascitic fluid is in small quantity and occupies
probably the pelvis only, the presence of dulness will scarcely be detected
in any ordinary position which the patient may assume: it may generally,
however, be certainly recognized if he be made to rest upon his elbows
and knees so as to allow the fluid to gravitate to the neighborhood of the
umbilicus. Thus, again, when peritoneal adhesions are present, both the

evidence derivable from the relative positions of resonance and dulness, and the variability of these positions, and that also derivable from fluctuation, may wholly fail us. Thus too when the abdomen is enormously distended, the attachment of the stomach and intestines may be too short to allow of any of these parts reaching the surface of the abdomen and the dulness may be universal, a condition which does not indeed throw any difficulty in the way of ascertaining the existence of fluid, but may make it not quite easy to determine whether the fluid is free in the abdominal cavity or whether it is contained in a large ovarian cyst. It need scarcely be said that, independently of the evidence afforded by the history of the case, by the form of the abdomen, and by vaginal examination, there is always in ovarian dropsy (unless indeed it be associated with Ascites) resonance in one or other or both flanks in consequence of the position which the tumor always takes in relation to the bowels; yet to insure accuracy it must not be forgotten that even in Ascites there may be a line of resonance in either flank due to the presence there of the colon. It must be added that œdema of the abdominal walls, or fat in them or in the mesentery, or the presence of diffused peritoneal cancer, are often serious impediments to the accurate diagnosis of moderate dropsical accumulations.

In most cases peritoneal dropsy causes merely that uniform distention of the abdomen which has been above described; but the distending force naturally exerts its most marked influence on those parts of the parietes which are weakest; and hence hernial sacs become often very greatly dilated and attenuated, especially perhaps the sacs of umbilical herniæ; hence, too, in some cases of Ascites in females the recto-vaginal pouch becomes greatly distended, and even protruded through the vulva in the form of a tumor, carrying with it as a covering the posterior wall of the vagina. I recollect one case in which the formation of such a tumor caused not only prolapse of the whole of the posterior wall of the vagina, but also of the upper part of the anterior wall together with the os uteri, which latter was found on the convexity of the tumor. The body of the uterus retained its normal position, but its neck had by the traction exerted on it by the gradual descent of the posterior wall of the vagina been attenuated and drawn out to a length of three or four inches. Occasionally Ascites has been relieved by the spontaneous rupture or perforation of some thinned portions of the abdominal parietes.

Œdema of the lower extremities and intervening parts is a very general and early accompaniment of abdominal dropsy. Sometimes it occurs at so early a period as to be the first symptom of disease which the patient himself recognizes, and indeed it is not very uncommon for ascitic patients to assert that their illness began with swelling of the legs. There is no doubt that in dropsy from abdominal disease this complication is due to the impediment to the return of blood produced by the pressure of the ascitic fluid on the iliac veins. It increases for the most part with the increase of the conditions on which it depends; and may become as excessive as that from cardiac or renal disease; but it rarely extends beyond the part with which the mechanically-impeded veins are immediately connected, and never becomes general. It need scarcely, however, be said, that when Ascites is connected with diseases of the heart, lungs, or kidneys, general anasarca is very often present. Anasarca due to abdominal dropsy is generally equal in the two lower limbs; and in this respect differs for the most part from anasarca in the legs resulting from abdominal tumors or from obstruction by clot of the iliac veins.

Shortness of breath is an early symptom, and it increases with the in-crease of the dropsy. It is not always noticed by the patient himself while he remains quiet in the sitting or semi-recumbent posture. But even at such times the physician will probably observe that the respiratory acts are unduly quick and shallow. Ultimately, however, this symptom becomes very painful and distressing. It is obviously caused by the en-croachment of the enlarging abdomen upon the thoracic cavity, by which the diaphragm becomes pushed up and prevented from performing the movements necessary for perfect respiration. The lower portions of the lungs become consequently more or less empty of air and collapsed; and, as might be anticipated from a knowledge of its cause, it is also much ag-gravated when the patient lies down.

Although in the earlier stages there may be little or no abdominal dis-comfort, there generally arises in the course of the affection a good deal of aching, which is usually complained of most in the flanks and across the epigastric or umbilical regions. This is probably due to the pressure which the fluid exerts on the various tissues, but more particularly to that which it exerts on the hollow viscera. This pain is sometimes associated with that of distinct colic, and not very infrequently, when the abdomen has become very largely distended, with pain of a peritonitic character. In-deed, acute or sub-acute peritonitis is far from rare in the latter stages of Ascites. It may be added, that diarrhoea is not uncommon in the course of Ascites, and that it seems to be sometimes due to the same impediment to the portal circulation which causes the Ascites itself, and is sometimes dependent on some slight dysenteric inflammation; and that although early in the affection there may be no visible morbid condition of tongue and neither thirst nor loss of appetite, the tongue and the digestive functions after a while all become variously and more or less seriously affected. It may be added further, that patients almost invariably complain of flatu-lency, a complaint which is undoubtedly due in many cases to excessive flatulent distention of the bowels, but may in some degree be explained by the discomfort which, in the presence of much ascitic fluid, even a nor-mal amount of gaseous distention may occasion. There is generally some dryness of skin and some diminution in the urinary secretion.

There are many symptoms, more or less grave, besides those which have been considered, which may be presented by ascitic patients; but they are symptoms for the most part due to the diseases upon which the Ascites itself depends, and are sufficiently considered elsewhere under the heads of those diseases.

Treatment.—The treatment of Ascites, in a large proportion of cases, merges in the treatment of the disease by which it has been caused. Still, in some cases from the very beginning, and in most when the accumula-tion becomes very great, special treatment directed against the Ascites is, or appears to be, called for. To promote the absorption and removal of the ascitic fluid there are good theoretical reasons for the employment of those remedial measures which increase the discharges from the skin, the kidneys, and the bowels. The skin in cases of Ascites is usually unnatu-rally dry, and this fact seems to furnish an additional argument in favor of the use of diaphoretics. There is no doubt, indeed, that diaphoretic remedies are very generally beneficial to the patient. And amongst these must not be forgotten the most powerful of all, namely, the hot-bath, the vapor-bath, and the Turkish bath. Again, the frequent diminu-tion in the urinary secretion may be urged as a further motive for the employment of diuretics; and again, it may be stated generally that the

promotion of the flow of urine is serviceable. Still more, the close connection between the peritoneal membrane and the mucous lining of the
bowels, and the fact that in hepatic obstructions the mesenteric capillaries
sometimes relieve themselves by discharge of serum at the serous surface,
sometimes by the escape of serum or blood at the mucous surface, would
seem to be decisive as to the value of purgatives, and more especially of
watery purgatives; and it may be freely admitted that purgatives are
very often beneficial. I must confess, however, that although fully ac-
quiescing in the importance of restoring as far as may be, and of main-
taining, the healthy action of the skin and kidneys, and of promoting a
tolerably free action of the bowels, I have never, to the best of my recollection and belief, seen an ascitic patient materially relieved as regards his
Ascites, far less cured, by a course of either diaphoretics, diuretics, or
purgatives. And in respect to purgatives, I may add that I have frequently had to discard them because, while they were not distinctly bene-
fiting the dropsy, they were obviously affecting the patient's health injuriously; and further, that according to my own experience, diarrhœa is a
not infrequent concomitant of Ascites, and is often difficult to arrest, and
often of bad augury. There are, however, certain medicines which are
more or less diuretic in their action which have been, or are, supposed to
have, occasionally at least, a specific influence over dropsical accumulations in serous membranes, and under the use of which occasional recoveries are recorded. Among these may be enumerated mercury, iodide and
bromide of potassium, copaiba, and the combination of fresh squills and
crude mercury.

But it must be repeated that, as a rule, the treatment which is
directed towards the alleviation or cure of the disease or condition of
health to which the Ascites is secondary, is that which is most likely to
be curative as regards the dropsy. The modes of treating heart diseases,
kidney diseases, bronchitis, cirrhosis, and so on, need not be here discussed;
but it may be pointed out that in a considerable number of cases of Ascites, and even in many of those in which the Ascites is dependent on the
diseases which have just been enumerated, there is present a greater or
less degree of anæmia and want of tone, and that in some at least of
these cases anæmia and want of tone are in some degree instrumental in
producing the dropsy. It is certain that tonics are very often well borne
by ascitic patients, and that even when not well borne at first a little
judiciousness in their employment, or in the employment of other prepara-
tory measures, will render them tolerable; and it is certain that under their
use ascitic patients do often not only improve in health, but lose, in part or
wholly, their dropsical accumulation, and that occasionally the recovery is
permanent, and permanent even after the performance of paracentesis.
Quinine, iron, and cod-liver oil are probably the most valuable forms of
tonics.

Counter-irritants and other forms of local applications are doubtless
sometimes useful for the relief of uneasiness and pain; but no such applications are of use in promoting absorption of the fluid. But when the
abdomen has become very much distended, and the patient is suffering
seriously from the inconvenience and distress which attend such distention, the removal of the dropsical fluid by paracentesis becomes necessary. The time for the performance of this operation must be determined
for each case, less by the actual distention of the belly than by the gravity of the symptoms which attend that distention. The operation is generally postponed as long as possible, and I believe rightly; but it may be

worth while to state that it has not very infrequently appeared to me that the beneficial effects of remedies have been exerted in a much greater degree after paracentesis than while the belly was largely distended. Paracentesis is generally a harmless operation; but sometimes peritonitis ensues, and is apt to be rapidly fatal. I believe that in cases of peritoneal dropsy dependent on cancerous disease of the abdomen tapping is not only very rarely of even temporary benefit, but that it generally hastens death. Iodine and other substances have occasionally been injected into the peritoneum for the cure of Ascites, and successful cases of this hazardous kind of treatment are recorded.

ABDOMINAL TUMORS.

By S. O. Habershon, M.D.

Before proceeding to the consideration of true abdominal tumors, it may be well to notice those which are of a delusive character, and have been called by Dr. Addison and Sir Wm. Gull *"phantom tumors."* It is a common thing for patients to suppose that there is something seriously wrong because one portion of the abdominal walls projects more prominently than another; sometimes the left hypochondrium is found to be enlarged from a flatulent stomach, or the cæcum or sigmoid flexure from similar gaseous distention, or the abdominal wall yields so as to form a direct protrusion; these conditions are easily recognized: the part is found to be flatulent on percussion, and manipulation fails to detect any solid growth; but in the "phantom tumor" a solid mass is felt, but it is in the parietes, and it is due to muscular contraction; the part is hard and dense, and may be readily mistaken; it is, however, fairly resonant on percussion, and by gentle and continued manipulation the muscle relaxes; if the hand be gently placed on the hard mass, and the attention of the patient diverted by conversation of an absorbing character, at the same time that the fingers are gently moved about the mass, the hardness disappears. This contraction of the muscular walls may be found at any part; sometimes it is on the right side, and the patient seems to have enlargement of the liver; frequently it is one of the transverse muscular bands of the rectus, sometimes it is the quadratus lumborum, or the transversalis muscle of the abdomen.

It is scarcely correct to speak of a loose kidney as a "phantom tumor;" it is a movable one, but it does not entirely disappear by pressure, although it may pass beyond the reach of the hand.

It is important in the study of abdominal tumors to have a definite acquaintance with the exact position of the abdominal viscera. For the convenience of description, the abdomen is divided into several regions marked out by lines from fixed points. A line drawn round the upper part of the abdomen at the most prominent part of the costal cartilages, and a second at the crest of the ileum, divide the surface into three zones, and these again by two perpendicular lines are subdivided each into three other spaces, the lines passing downwards from the cartilage of the eighth rib to the middle of Poupart's ligament. In the upper zones we have the right and left hypochondriac regions, on either side of the epigastric space or scrobiculus cordis; in the central zone, the right and left lumbar regions are on either side of the umbilical, and below, in the third zone, the right and left iliac are situated on either side of the hypogastric region.

During health these regions are occupied by their respective viscera, and
it is useless to try and ascertain abnormal states, unless there be a thorough
knowledge of that which is normal. We would further add that in every
case of abdominal tumor it is important to enquire: 1st, into the general
history of the symptoms of the patient; 2d, to ascertain the exact position
of the tumor, and the physical signs; and 3d, to learn whether there is
any functional disturbance of the abdominal organs. It is not likely that
any viscus is involved in a morbid growth, if it perform its functions in a
healthy manner.

It may be well to consider these regions as regards their normal and
abnormal contents. The *right hypochondrium* contains especially the
liver and gall-bladder, but the gland passes into the epigastric region, and
reaches the left hypochondrium; when enlarged it extends into the umbilical and right lumbar regions. The liver is attached to the diaphragm,
and to a certain extent moves with it. It reaches upwards as high as the
fifth rib, where the dulness commences and is partial; at the sixth rib the
dulness is complete; when the patient is recumbent the liver is behind the
ribs, unless, as is generally the case in women, it has been pushed down
by compression; but in the erect and sitting postures the liver may be felt
an inch below the ribs. Beside the liver and gall-bladder the right hypochondrium contains the angle of the ascending colon, part of the duodenum, the right supra-renal capsule, and the upper part of the right kidney. It must be remembered, in reference to enlargement of the liver, that
the gland may be *pushed* down by *pleuritic effusion.* In large effusions
into the right pleura the liver is always displaced.

2. The liver may be pushed down by effusion between the upper lobe
and the diaphragm, the effusion being either serous or purulent; in these
cases the symptoms may closely resemble pleuritic effusion. Many years
ago a woman was admitted under my care into Guy's Hospital with severe
peritoneal symptoms after a fall from a cart upon the abdomen. The
pain was great, and the strength gradually gave way. On examination
of the lower part of the right lung before death, dulness on percussion,
with bronchial breathing and modified voice sound, was heard, and some
who had not known the previous history believed the disease to be above
the diaphragm. On the post-mortem table a trochar was introduced between the ribs posteriorly, and pus exuded, apparently confirming the idea
that empyema existed; but on fuller examination it was found that the
pus was situated between the liver and the diaphragm; there had been
peritonitis, and the pus was circumscribed by adhesions.

3. The liver may be pushed down by the development of tumors, or
by a hydatid cyst in the right lobe of the liver. This cyst or tumor may
be so situated behind the ribs as to be quite beyond the reach of the hand.
A woman in middle life was admitted under my care into Guy's with the
liver extending into the umbilical region, the surface was smooth, the
gland passed lower into the abdomen, and the strength of the patient at
length gave way and she sank. On examination, when the abdomen was
opened, the liver seemed to fill the greater part of the abdomen on the
right side; the surface was smooth, and it was only when the gland was
drawn down from the diaphragm that the true nature of the enlargement
was recognized. An enormous hydatid cyst occupied the right lobe and
pressed in every direction, but did not reach the free surface of the liver.
In the examination of an enlarged liver the patient should be placed on
the back, the knees drawn up, and the head comfortably supported, and
the patient's attention should be absorbed by conversation, if possible.

Sudden pressure will often detect an enlarged gland, which could not be recognized by gradual pressure; if the fingers be passed upwards from below, the edge may be caught, and they should then be gently passed over the gland to ascertain whether there are any irregularities; the surface is dull on' percussion; at the upper part where the liver is over-lapped by lung there is partial resonance, and we sometimes find that the lower edge may be covered by the colon, and resonance is thus produced.

4. Enlargement of the gall-bladder is found generally opposite the tenth rib; it is pyriform, and when filled with bile will yield somewhat to pressure. Sometimes I have found it filled with a great number of gall-stones, so that it resembled a solid mass; a very different condition was found some years ago in a case under the late Dr. Babington, the specimen of which is in Guy's Museum; the gall-bladder communicated with the intestine, and was filled with gas, so that there was resonance on percussion.

5. Enlargement of the liver, if general, arises from congestion, from inflammation, from fatty deposit, from lardaceous disease, from obstruction of the bile-ducts ; if *local*, from hydatid tumor, from syphilitic deposit, from cancerous growths, from abscess in the liver, or from suppuration in connection with the bile-ducts.

6. An enlargement of the right kidney can sometimes be felt immediately below the right lobe of the liver; it will be found to extend into the loin.

7. Malignant disease of the right supra-renal capsule may also reach the lower part of the liver. I have found a tumor of the left supra-renal capsule simulating enlargement of the spleen.

8. Malignant disease of the pancreas and first part of the duodenum are recognized by their clinical history; the hardness may be felt closely in contact with the liver.

9. Cancerous disease of the angle of the ascending colon is also felt in this region; in this disease pain comes on several hours after food, and there is likely to be discharge of blood or of mucus from the bowels.

In the *epigastric* region we find the stomach and its lesser curvature, and the left lobe of the liver; the gall-bladder and the pyloric extremity of the stomach are situated at its union with the right hypochondriac region; posteriorly we have the pancreas, the aorta, the vena cava, the cœliac axis and the commencement of its large branches; at the lower part of this region we have the transverse colon, varying, however, in position according to the distention of the stomach. We would remark that yielding of the parietes, with flatulent distention of the stomach, often gives rise to the idea of tumor.

2. Abscess sometimes forms in the parietes in this part; not only in the muscular parietes, but in the loose cellular tissue about the end of the sternum. In a case of that kind, in the clinical ward of Guy's some years ago, there was a projection at the scrobiculus cordis, which was afterwards found to be an abscess which extended to the under surface of the diaphragm.

3. Abnormal pulsation is often felt at the epigastric region; this may arise from an aneurismal tumor in connection with the aorta or with the cœliac axis; in aneurism of the aorta close to the diaphragm it is, however, very difficult to feel the aneurismal tumor, unless it be of large size. In the majority of cases a pulsating tumor at the scrobiculus cordis is found to arise from disease of the left lobe of the liver pressing upon the

abdominal aorta; in other cases the pulsating mass may consist of a vascular and pulsating medullary growth in the stomach.

4. Tumors of different kinds in the left lobe of the liver are found in this region.

5. Disease at the lesser curvature of the stomach may not only be felt in this region, but may have pulsation communicated to it from the aorta.

6. Chronic ulcer with thickened walls may constitute the tumor felt at the epigastric region; but ·

7. A hard mass felt in the epigastrium is more frequently found to be malignant disease.

On a level with the umbilical line and situated posteriorly we have the *pancreas*, but so deeply is the gland placed that it is difficult to recognize enlargement by digital examination.

Malignant disease of the pancreas, where there is much enlargement, may be sometimes recognized; but in such cases we principally depend upon the clinical history and the general symptoms, in forming a correct diagnosis. In a case of inflammation of the cellular tissue about the pancreas, which I saw in consultation some years ago, there was a hard swelling felt at the upper part of this region; there was severe pain, with febrile excitement, and I supposed the case was one of gastric disease; post-mortem examination showed that the stomach was healthy, but that the swelling was an abscess connected with the pancreas.

In chronic disease of the omentum, often of a malignant character, the serous membrane is puckered and drawn upwards, so that it forms a firm band passing across the abdomen, at the lower part of the epigastric or at the upper part of the umbilical region. It may be associated with malignant disease of the peritoneum, as we have mentioned.

In the *left hypochondrium* we have the cardiac extremity of the stomach, the spleen, the left supra-renal capsule, and a portion of the kidney. The left angle of the colon may also extend to this space. The most common tumor found in this region is the spleen, which, as it increases in size, not only passes upwards and raises the level of dulness on that side, but it passes downwards and is also directed forwards. This forward direction may be due to the band of peritoneum immediately beneath the gland; but there is no doubt of the fact that the spleen, as it increases, and it attains sometimes an enormous size, passes not only into the lumbar but also into the umbilical region. The fissure at the anterior edge assists us in the recognition of the spleen; but this is not a certain sign, for the fissure between two enormously enlarged lymphatic glands may communicate to the touch the same impression. The spleen enlarges after a full meal; it may increase from temporary portal congestion; it is felt below its normal position in enteric and other fevers, but these conditions would not be designated as tumors. It is in the enlargement after ague, in leucocythemia, in lardaceous disease, that we find the spleen to attain to very large proportions, and in these cases the increase is general. In abscess, in hydatids, and in malignant disease, the enlargement is partial.

2. Another tumor in the left hypochondrium arises from enlargement of the left kidney or from hydatid at that part; but in this case the growth extends more into the loin, and there is more likelihood of distention of the lower ribs on that side.

3. Malignant disease of the supra-renal capsule on the left side sometimes presses forward to the anterior part, and is felt immediately below the spleen. I have known it mistaken for an enlargement of the spleen

itself, for the growth was large, and it appeared to pass from beneath the ribs.

4. Aneurismal disease at the commencement of the descending aorta sometimes pushes forward the spleen, and the diagnosis is difficult. In the case to which I refer, the lower ribs were prominent, but pulsation was very indistinct. The patient died from rupture of the aneurismal sac behind the peritoneum.

5. The solid walls of an ovarian cyst sometimes reach into the left hypochondrium, form a tumor, and simulate disease of the spleen. The tapping of the cyst would be one means of diagnosis of a case of this kind, for the solid portion of the cyst would then pass towards the pelvis.

6. It is scarcely necessary to mention the manner in which the spleen may be pushed down by effusions both into the pleura and into the pericardium.

7. Local suppuration may occur in this region, and produce a swelling resembling a tumor.

The next zone of the abdomen is divided into the central, the *umbilical*, and into the *right* and *left lumbar* spaces. In the *lumbar* regions the kidneys and the ascending and descending colon may be causes of tumors. The kidneys extend to the loin, and as they pass forward can be felt anteriorly to the inner side of the colon. The condition of the urine, in the discharge of blood, of pus, or of cancerous cells, affords to us an important guide in diagnosis. A tumor from distended pelvis of the kidney is most uncertain as to its size. Sometimes a calculus may block up the ureter, and the pelvis of the kidney gradually attains large dimensions, till at length the ureter is distended beyond the size of the obstructing calculus, and a sudden discharge of several pints of urine at once diminishes the size of the tumor. The same may be the case in suppuration of the kidney. In a patient under my care in Guy's, a large tumor extended from the right lumbar into the iliac region, and would suddenly subside on the discharge of several pints of urine. This condition came on when the patient was about sixteen; he was a shoemaker, and although he had several severe attacks, he continued his work till he was nearly sixty-four years of age, when he died in Guy's Hospital from a large cancerous growth which affected the kidney on the same side. After death it was found that a calculus was the cause of the obstruction of the ureter.

2. The glands in connection with the kidney, the lumbar glands, sometimes form a large tumor in this region; the urine is then unaffected, the growth is more irregular, and there are generally other indications of malignant disease. In hydatid disease of the kidney the tumor is rounded and elastic.

3. Accumulations in the ascending or descending colon form masses in these regions; but the clinical history, the absence of severe symptoms, and the relief by purgative medicine, characterize these cases.

4. In intussusception a doughy, elongated mass may be felt in the ascending or even in the descending colon. The severe and spasmodic pain, the vomiting which is often present, the obstruction of the bowels, and the discharge of blood and of mucus, indicate the character of the affection.

5. In diseases of the spine leading to suppuration there is a bulging in the loin, and it may be a projection anteriorly, which can be felt on digital examination.

6. Abscess in the loin and in the quadratus lumborum muscle may lead to enlargement, and may extend into the bowel. Thus, we have known an abscess pass into the cæcum, and have seen suppuration primarily connected with the bowel reach the loin.

7. Ovarian tumors may pass from the iliac fossa into the loin, but more frequently they extend into the umbilical region.

In the *umbilical* region we have the transverse colon and the omentum attached to it, the small intestine, with the mesentery and mesenteric glands, and posteriorly the aorta and vena cava. On either side are the right and left renal vessels. The position of the transverse colon varies greatly; sometimes its curve is greatly increased, and it may reach nearly to the hypogastric region. The curve may be increased as the consequence of distention, or by the dragging of an omental hernia. Tumor in the walls of the intestine also changes its position.

2. Intussusception of the small intestine is often found in the umbilical region.

3. Tumors in the omentum and in the mesentery. In the former, the mass is movable, and there is no functional disturbance of the intestine.

4. In strumous disease of the intestine the bowels are often matted together by inflammatory adhesion, and the mass resembles a tumor. Sometimes there is suppuration or fæcal abscess, or it may be the discharge takes place from the umbilicus.

5. Ovarian tumors are often found to extend into the umbilical region.

6. Aneurismal disease of the aorta and the branches of the abdominal aorta may be felt in this space, but frequently enlarged glands, pressing upon the aorta or upon the renal arteries, simulate true aneurismal disease. Sometimes we can remove the gland from the pulsating vessel beneath, by manipulation or by changing the position of the patient; but this is not invariably the case.

In studying enlargement in the loins, it is important to remember that both the ascending and descending colon closely sympathize with diseased structures before or behind them, the intestine becomes inactive, the passage of the contents impeded, and the fulness may resemble primary tumor of the bowel.

The remaining regions are the *hypogastric* and the right and left iliac.

In the *hypogastric* region whenever a tumor is felt we must always render ourselves sure that the bladder is not distended. Urine may pass constantly, in fact, there may be a constant dribbling from over-distention of the bladder, and thus the disease be unsuspected. In this way I have known the bladder reach above the umbilicus and contain many pints of urine.

2. Enlargements of the uterus, whether from pregnancy or tumor, extend into the hypogastric region.

3. Fibroid tumors of the uterus and of the ovaries, and cystiform disease of the ovaries, extend into right or left iliac region, and also into the central space.

4. Hydatid disease of the cellular tissue in connection with the bladder forms a rounded tumor in this space, in touch very closely resembling a distended bladder.

5. Aneurismal disease of the iliac vessels must also be borne in mind as a cause of tumor in the lateral portions of the hypogastric region.

In the *right iliac* region we have the cæcum and its appendix, and diseases of these structures constitute many of the morbid enlargements at this part. The mischief may, however, be external to the cæcum, peri-typhlitis; in these diseases we have local pain and tenderness, febrile excitement, generally a disordered condition of the bowels, constipation, sickness, and it may be peritonitis. In tumor from enlarged glands, there is less interference with the action of the bowels. In pelvic abscess the bowel is free, the mischief is found to extend from above, and it passes onwards in the direction of the psoas and iliacus muscles. A tumor from renal disease may reach the iliac fossa, but it can be traced upwards, and the urine will be found to be diseased in most cases, if carefully examined, except in instances in which one kidney does the entire work and the other is completely shut off.

An aneurismal tumor from disease of the iliac vessels is recognized by its pulsatile and expansive character, and by the condition of the circulation of the limb.

Ovarian tumors reach to the right or left loin; sometimes they become adherent to the bowel, and I have known an ovarian cyst become adherent to the cæcum, and having discharged its contents into the bowel, the cyst has become filled with fæcal matter. Acute disease of the right ovary sometimes closely resembles typhlitis, but the pain is lower, it extends into the pelvis; the tenderness, the constipation, the febrile excitement, may be equally marked in the acute disease of the ovary, as in cæcal disease.

In the *left iliac* region many morbid growths correspond to those on the right side, but here the sigmoid flexure takes the place of the cæcum. The curvature of this part of the bowel varies greatly; sometimes the sigmoid flexure extends to the right side, and there may be adhesion to the cæcum. In other cases of distention it bends upon itself, it falls into the pelvis, and the acute bending at a right angle leads to obstruction. The termination of the sigmoid flexure in the rectum at the brim of the pelvis is the part often affected by disease, and a tumor can in most cases be made out by careful manipulation. Sometimes there is a rounded growth in connection with the mucous membrane of the bowel, at other times all the coats are thickened and contracted, as if a piece of string had been tied round the bowel; all these states lead to gradually increasing obstruction, which may become complete.

Inflammatory adhesion sometimes takes place between the sigmoid flexure and the bladder; an external tumor is felt on deep but gentle manipulation. The diagnosis and the prognosis of these cases are often very obscure, and in several we have known direct communication take place with the bladder, and fæcal discharge with the urine supervene. Colotomy is of the greatest service in these cases.

We have thus briefly sketched the site and the character of abdominal tumors; each case has a clinical history of its own. and it is by the careful study of that history, in connection with the position of the tumor and the disturbances of the functional activity that are associated with it, that we can make out the true nature of the disease. Many most interesting cases of abdominal tumor might have been added to this chapter; the difficulties in the diagnosis might thereby have been more fully indicated, and the various modes of relief discussed; but we have refrained on account of the length to which this work has already extended, and we have only given the general facts which these instances of disease have brought out.

We have sought to show the leading characteristics of diseased conditions as manifested in the various portions of the alimentary canal; and
to do this have recorded the cases themselves, as facts upon which each
one may form his own opinion, rather than depend entirely upon the de-
ductions we have drawn from them. Such general conclusions in most
chapters have preceded the cases upon which they are founded; and we
leave them before our readers with the hope that they will serve further
to elucidate the general symptoms, pathology, and treatment of diseases
of the alimentary canal.

INDEX.

www.ingramcontent.com/pod-product-compliance
Lightning Source LLC
Chambersburg PA
CBHW021525210326
41599CB00012B/1389